EVOLVING EDEN

EVOLVING EDEN

An Illustrated Guide to the Evolution of
the African Large-Mammal Fauna

ALAN TURNER AND MAURICIO ANTÓN

Columbia University Press New York

Columbia University Press

Publishers Since 1893

New York   Chichester, West Sussex

Library of Congress Cataloging-in-Publication Data

Turner, Alan, 1947–

Evolving Eden : an illustrated guide to the evolution of the African large-mammal fauna /

Alan Turner and Mauricio Antón.

p. cm.

Includes bibliographical references (p.).

ISBN 0-231-11944-5 (cloth : alk. paper)

1. Mammals—Evolution—Africa. 2. Mammals—Evolution—Africa—Pictorial works.

I. Antón, Mauricio. II. Title.

QL731.A1T87 2004

599.138′096—dc22

                                                    2003068832

Designed by Kristina Kachele Design LLC

Composition by A. W. Bennett Inc.

Columbia University Press books are printed on permanent and durable acid-free paper.

Printed in the United States of America

c 10 9 8 7 6 5 4 3 2 1

FRONTISPIECE: *Australopithecus africanus* at Swartkrans fleeing from a saber-toothed cat of the genus *Homotherium*.

For Adam and Miguel
Puri and Gill

CONTENTS

Foreword

THE TOPIC OF THIS BOOK is exciting, scientifically important, and timely. Africa has a unique significance for understanding the evolution of the larger terrestrial mammals, whose prolific radiation on the African savanna during the period addressed in this book—the past 35 million years—exceeded that on other continents. This diffusion was influenced by Africa's position as the largest landmass straddling the equator. Africa is also the continent on which the past mammalian diversity has survived to the greatest extent. This African combination—as a past "species factory" and a present foremost "museum" that preserves much of that diversity—is particularly favorable for an understanding of the nature of mammalian, including human, evolution.

New information on this subject has appeared at an exponential rate over the past decades, while synthesis has lagged behind. *Evolving Eden: An Illustrated Guide to the Evolution of the African Large-Mammal Fauna* is a timely work in that Alan Turner and Mauricio Antón lucidly introduce many of these diverse additional findings and show their relation to one another and to previous knowledge. Some of the new evidence is on the African mammals themselves in the form of numerous descriptions and analyses of fossil species, including early human relatives, and of fossil assemblages. There also has been an upsurge of interpretations of how extinct and extant species relate to one another on the mammalian phylogenetic tree. Such evidence, together with new physical dates for fossil occurrences, has served to place ancient evolutionary events in phylogenetic and chronological context. An even broader background has been provided by the burgeoning number of studies of past environmental changes (including those in global and local climate, in sea level, and in Earth's crust, such as continental drift, mountain uplift, and volcanism) and their effects on African ecosystems and lineages. In addition, lively recent and ongoing debates touch on such evolutionary questions as the appearance and extinction of species and the relation of species' turnover (including speciation and extinction) to ecosystem formation and disappearance. How do changes in organismal form and function evolve, how do they relate to speciation and ecology; and how do events in the evolution of humans relate to those in the evolution of other mammalian lineages? Even for the expert, it is difficult to put together all these heterogeneous strands of knowledge into a coherent whole because they are distributed among so many disciplines and literature sources. This book is remarkably successful not only in presenting many of these basic pieces of evidence but also in integrating them into an accessible synthesis.

The first requirement of such a synthesis—to bring back to life the extinct animals in terms of what is known about their morphologies, lifestyles, and communities—is one of the many aspects of this book that is exceptionally well done. Both authors are experienced specialists on mammalian anatomy. For decades, Alan Turner has been a highly respected systematist of carnivore fossils and analyst of large-mammal communities and evolution. Mauricio Antón is a gifted artist and foremost scientific illustrator. His skill and expertise as an anatomist are evident in the more than 100 exquisite and anatomically accurate illustrations that reconstruct extinct mammals, scenes from the communities in which they coexisted, and the landscapes in which they lived. I do not know of any published mammalian reconstructions that are better than these. They are based directly on the fossil skeletons and skulls of the species being reconstructed. Together, the illustrations and text bring back to life the extinct animals with scientific integrity, and do so as well as has ever been done.

Throughout this book, Turner and Antón have not avoided specialized scientific subjects just because it might be difficult to include them. Instead, they present each topic, from particulars of morphology to those of community structure and climate change, in a form that is readily accessible while maintaining scientific accuracy. Similarly, the authors have risen to the challenge of presenting some of the more complex evolutionary issues, such as the current unresolved debates on the causes of speciation and the role of physical changes in initiating species' turnover events. It is an added bonus of this book that it clearly and simply introduces the reader to this and other continuing discussions in modern evolutionary theory.

Today there is a renascent acknowledgment in paleontology that understanding our place in the biosphere requires an analysis of the Hominidae—the family to which we belong—in the context of the other lineages that evolved alongside ours. Charles Darwin realized this. He wrote, for example, in a notebook dated 1837 that "man is a species like any other. The mind is a function of body. He who understands baboons would do more towards metaphysics than Locke." But somehow that integrative perspective took a backseat to a narrower focus on the hominid fossils for much of the century and a half since his time. Turner was one of the pioneers of a renewed synthetic approach to the evolution of hominids in the context of mammalian communities, with his examination in the early 1980s of the many intercontinental migrations of early humans. He showed repeated patterns of human movements as part of general mammalian dispersions, with fellow travelers that included particular large carnivore species as well as other mammals. This contextual perspective is very much in evidence in this book: the reader who wants to know which other animals lived and interacted with members of our lineage at different times and in different localities during the past 35 million years will find not only the basic information but also analyses of what was happening among the different species—with the beautiful and carefully researched reconstructions of both extinct and extant forms in ancient natural settings making the whole story instantly real.

In sum, this very enjoyable volume is full of fascinating information and rich in synthesis. It is an excellent book for the nonspecialist and the lay public because of an engaging writing style that is well integrated with abundant and beautiful illustrations. At the same time, Turner and Antón, with the unique combination of talents and experience that they bring to this book, have taken pains to remain scientifically accurate. They have not shied away from presenting a very broad range of information, including important issues in modern evolutionary theory. Because of this integration of numerous disparate bodies of material, as well as a bibliography—organized by topic—with suggestions for more detailed reading, specialist colleagues and paleontology students also will find it a useful and enjoyable book. I know that I will. *Evolving Eden* deserves a wide readership and is not to be missed by anyone interested in our origins.

Elisabeth Vrba
Department of Geology and Geophysics
Yale University

PREFACE

IT HAS LONG BEEN APPARENT that human origins lie in Africa, as the wealth of fossil evidence found there at an ever-increasing pace during the twentieth century amply demonstrates. We can trace our own immediate lineage, the genus *Homo*, back at least 2 million years (Ma). We can see evidence of ourselves and our closest fossil relatives—members of the genera *Australopithecus, Kenyanthropus, Ardipithecus,* and *Paranthropus**—for perhaps 2.5 Ma before that. We can date yet more remote ancestors in Africa to around 35 Ma, the time when the Primates, the order to which we and both the apes and the monkeys belong, first entered the continent.

We humans are, in short, just like all other animals—the current end point in a long, long line that stretches back into the deep past—and logically we have no single point of origin in either time or space. And yet the metaphor of an Eden of creation for human and other life persists, despite the scientific understanding of evolution and the direct evidence of the fossil record. When suggestions that fully modern humans originated in and dispersed from Africa about 100,000 years ago began to appear in the scientific literature during the 1980s, the popular press latched onto the notion of an "African Eve" as the last common female ancestor of us all. An African Eve, with or without an Adam, implies an African Eden.

The continent of Africa is, of course, a reasonable candidate for description as a Garden of Eden. Now set astride the equator—although it has moved northward during the 35 Ma of primate presence—it has generally been warm, richly vegetated, and

---

*These names, and the use of scientific names for species in general, are discussed in chapter 3.

stocked with a large number of animals of all shapes and sizes. But if Africa is to serve as a modern-day version of Eden, we must understand that the continent itself has not remained unchanged—particularly in the time since the first primates appeared there. Africa, like all other landmasses, has undergone profound alterations in its physical geography, climate, and biota over millions of years. Eden, in other words, has evolved.

Evolving Eden examines the evolution of the mammalian fauna of the African Eden against changes in the topography, vegetation, and climate of the continent. The 35 Ma or so of primate presence makes a rather convenient period over which to tell this story, since much happens and larger patterns thus may be discerned. Because dealing with the evolution of the mammalian fauna of a whole continent in one volume is a rather tall order, we have chosen to restrict ourselves by examining essentially the larger mammals, although our smaller relatives are also mentioned in order to provide a more comprehensive picture.

The structure of this book is designed to link the evidence of past and present. Chapter 1 introduces such general topics as dating, continental drift, global climate change, and the likely motor of evolution, while chapter 2 provides an overview of the physical evolution of the African continent, its present and past climates, and the major determinants of plant and mammal distribution. This background is followed in chapter 3 by an explanation of the methods of reconstructing extinct species, a discussion of the terminology of taxonomy, and a summary of the fossil history and modern distribution of each of the terrestrial mammalian orders known from Africa, together with an overview of the main features of each family within the orders. The extensive illustrations of living and fossil species in the chapter both enhance the text and bring to life the story of evolution in Africa. The main fossil sites in Africa and a summary of their important features—such as age, environmental reconstruction, and range of fauna—together with some depictions of key localities during the time of deposit formation, are the subject of chapter 4. Finally, chapter 5 draws together the information presented in the first four chapters to provide a summary of the evolution of the African mammalian fauna, including that of the human lineage, against this larger background, and illustrates groups of some of the more important species of larger mammals to give a sense of scale.

Our aim is to enable the reader to dip into the book in various places in order to obtain both detail and overview and, in the process, to present human evolution as very much a part of the larger pattern, while responding to the unique interest that we modern representatives of the family Hominidae have in our own past. For those interested in more detail than we can provide on any one aspect of this huge subject, the book concludes with a bibliography organized by topics. The bibliography is extensive, but it is also selective rather than exhaustive; many of the references offer good starting points for those who wish to pursue particular subjects, but the available literature on many groups, especially our own ancestors, is simply enormous.

A Note on the Photographs

All the photographs of extant African mammals and their environments in this book were taken by the authors, except for some landscape shots that were kindly provided by friends and colleagues, as specified in the captions. The pictures are of free-ranging animals, not zoo specimens. We think that this choice serves to emphasize that Africa is the one continent on Earth where one need not be a professional photographer and/or a field biologist to encounter such a variety of wild mammals and to capture them on film.

Seeing lions, rhinos, and mammoths was an experience of everyday life for Paleolithic Europeans, although it filled them with fresh wonder that they expressed in the cave paintings of Chauvet, Lascaux, and Altamira, to mention the most famous examples. But what remains of the European fauna is so widely scattered, and so scared of humans, that we can hardly expect to see many wild animals in the countryside. The situation is only a little better in the Americas, Asia, and Australia. That Africa can still offer such impressive visual experiences to the casual tourist is as much a testimonial to the work of African conservation authorities as it is to the low impact of Pleistocene extinctions. A photographic safari to Africa is an affordable experience for many middle-class Westerners, at least once in a lifetime. With all its debatable aspects, eco-tourism is one of the most straightforward ways for us as individuals to contribute to the economies of African countries and the conservation of African wildlife.

ACKNOWLEDGMENTS

WE ARE GRATEFUL to many people for assistance, discussion, and access over many years to material held in museums in Africa, Europe, and North America. In particular, we wish to thank Jordi Agustí and Salvador Moya-Sola, Institut de Paleontología Miquel Crusafont (Sabadell) • Luis Alcalá, Museo Nacional de Ciencias Naturales (Madrid) • Angel Galobart, Museo Arqueológic Comarcal de Banyoles (Banyoles) • Judy Maguire, Bernard Price Institute for Palaeontological Research (Johannesburg) • Germaine Petter, Muséum Nationale d'Histoire Naturelle (Paris) • Michel Philippe, Musée Guimet d'Histoire Naturelle (Lyon) • Abel Prier and Roland Ballesio, Université Claude Bernard (Lyon) • Marie-Françoise Bonifay, Laboratoire du Quaternaire, CNRS Luminy (Marseille) • Inesa Vislobokova, Paleontological Institute of the Russian Academy of Sciences (Moscow) • Marina Sotnikova, Geological Institute of the Russian Academy of Sciences (Moscow) • Lorenzo Rook, Paul Mazza, and Frederico Masini, Universitá degli Studi (Florence) • Peter Andrews, Andrew Currant, and Jerry Hooker, The Natural History Museum (London) • Brett Hendey, South African Museum (Cape Town) • David Wolhuter and Francis Thackeray, Transvaal Museum (Pretoria) • Meave Leakey, National Museums of Kenya (Nairobi) • Hans-Dietrich Kalhke, Ralf-Dietrich Kahlke, and Lutz Maul, Institut für Quartärpaläontologie (Weimar) • Jens Franzen, Forschungsinstut Senckenberg (Frankfurt) • Adrian Friday, Zoological Museum (Cambridge) • and André Keyser, University of the Witwatersrand (Johannesburg).

We also owe a great debt to many friends, colleagues, and collaborators for information, photographs, discussion, advice, encouragement, and hospitality, especially

Dan Adams, Emiliano Aguirre, Jordi Agustí, Luis Alcalá, Peter Andrews, Juan-Luis Arsuaga, Rolland Ballesio, Jon Baskin, Gérard de Beaumont, Lee Berger, José-Maria Bermudez de Castro, Raymond Bernor, Laura Bishop, Bob Brain, Tim Bromage, Ron Clarke, Andrew Currant, Eric Delson, George Denton, Yolanda Fernández-Jalvo, Giovanni Ficcarelli, Robert Foley, Ann Forsten, Mikael Fortelius, Susana Fraile, Angel Galobart, Nuria Garcia, Rosa García, Léonard Ginsburg, Francisco Goin, John Harris, Gary Haynes, Brett Hendey, Clark Howell, Nina Jablonsky, Peter Jackson, Christine Janis, Derek Joubert, Ralf-Dietrich Kahlke, André Keyser, Kathy Kuman, the late Björn Kurtén, Angela Lamb, Louise Leakey, Meave Leakey, Margaret Lewis, Adrian Lister, Martin Lockley, Gregori López, Larry Martin, Ignacio Martinez, Jay Matternes, Colin Menter, Gus Mills, Clare Milsom, Plinio Montoya, Jorge Morales, Mary Muungu, Manuel Nieto, Jim Ohman, Francisco Pastor, Charles Peters, Germaine Petter, Martin Pickford, Manuel Salesa, Israel Sánchez, Robert Santamaría, José Luis Sanz, Friedemann Schrenk, Chris Shaw, Andrei Sher, Nancy Sikes, Dolores Soria, Marina Sotnikova, Fred Spoor, Anthony Stuart, Richard Tedford, Phillip Tobias, Danilo Torre, Jan Van der Made, Blaire Van Valkenburg, Elisabeth Vrba, Lars Werdelin, Eleanor Weston, Peter Wheeler, Alisa Winkler, and Bernard Wood.

Among those, we especially thank Meave Leakey, who provided firsthand information essential for several of the reconstructions of East African fossil mammals in this book, and Francisco Pastor (Facultad de Medicina, Universidad de Valladolid), who carried out the dissections of several mammals that served as the bases for many of the reconstructions. We also thank Christine Janis and Lars Werdelin for constructive and helpful reviews of the text and illustrations that reduced errors and ambiguities. Those that remain are our own fault.

At Columbia University Press, we thank Alessandro Angelini, Holly Hodder, and Robin Smith for their help and support, and Irene Pavitt for her excellent editorial work.

Finally, it is our pleasure to thank Elisabeth Vrba for contributing a foreword to this book. Her work on the relationship of changes in the physical environment to evolutionary events in the biota, and particularly on the evolution of African antelopes, has shed light on the whole subject of evolutionary motors and been of immense importance to our understanding of the patterning seen in the fossil record.

EVOLVING EDEN

1 Dating, Continental Drift, Climate Change, and the Motor of Evolution

AFRICA IS HOME TO MORE THAN 1100 mammal species, or around one-quarter of the total diversity of living mammalian life. Such is the variety of the mammalian fauna of Africa that it is only approached, but not equaled, by that of Eurasia and North America combined. Recent molecular analyses even suggest that a small number of mammalian orders—among them elephants, aardvarks, and elephant shrews—can be identified as a unique group, the Afrotheria (literally, African beasts), members of which have a restricted common ancestry. Yet despite its unique aspects, the mammalian biota of Africa shares features of its composition and structure with that of adjacent continents, as does the flora. These similarities stem not only from the present position of Africa but also from the evolutionary histories of the continental biotas and the shifting patterns of biogeography induced by continental drift and climate change.

Timescale: Dating and Correlation

Life has existed on Earth for more than 3.5 billion years, although the pace of change has quickened in the past few hundred million. Mammals, the warm-blooded class of animals to which we belong, are known for the past 250 Ma, but for the first 190 Ma or so of that time they coexisted with, and were seemingly dominated by, the dinosaurs. The true age of mammals really started only around 65 Ma ago, when the dinosaurs became extinct, and it is only since then that the diverse array of modern mammal body sizes and body types has emerged in a burst of what is often termed adaptive radiation. The rate of change over that time has been quite phenomenal and may be summed up by one observation. If you traveled back in time no more than, say, 5 Ma, anywhere on Earth, you probably would find no living species of mammal, with the possible exception of the African rhinos. You would, of course, find animals that looked broadly similar to many extant forms and were closely related to, and perhaps even direct ancestors of, familiar modern species. Moreover, the rate of change in some lineages has been greater than in others, for reasons that we shall return to at the end of this chapter. But the change over that relatively short period of time (short at least in geological terms) has been dramatic.

One more specific example from the relatively recent past can highlight the extent of change that must be understood when we examine the fossil record. Spotted hyenas, now confined to sub-Saharan Africa and thought of as a typically African species, roamed Europe and much of Asia until perhaps 20,000 years ago. In Eurasia, they hunted and ate reindeer, animals now confined to sub-Arctic areas, in a region where vegetation typical of steppe and tundra replaced deciduous forest and other types found in temperate zones and the rest of the fauna included such animals as bison, mammoths, woolly rhinos, and wolves. In other words, the predator–prey relationship and the entire environment in which it took place were totally different from those of the present day.

In large part, such changes in the distribution of animals and plants have been produced by massive alterations in climate. Over the very long term, the 3.5 billion years of the fossil record, climate change itself has been influenced to a great extent by continental drift. The movements of the continents have opened and closed seaways and have lifted up mountain chains that subsequently eroded away. In other words, every feature of the physical and biotic environment, including the positions of the continents and their points of contact, has changed over time. Time is therefore one of the most important things for the paleontologist to understand. In everyday terms, we think of time in years, perhaps in centuries, and, at most, in millennia. Geologists measure time in millions of years; our own ancestors can be traced back fairly directly for around 4.5 Ma. We are concerned here, in broad terms, with the past 65 Ma of the evolution of life, with attention focusing on the final 35 Ma of that span.

For the past 200 years or so, geologists and paleontologists have understood how to date sediments and the fossils found in them in relative terms, based on the detailed study of stratigraphy—the sequence of rock layers, or strata. Local sequences have been built up into regional and continental syntheses, with fossils used as a means of correlating deposits of similar age over vast distances and with names given to various periods of time dated by particular groups of fossils. This approach is known as a biostratigraphy and was the basis of the original identification, naming, and plac-

TABLE 1.1

Epochs of the Cenozoic Era

LOWER BOUNDARY (MA)	EPOCH
1.8	Pleistocene
5.2	Pliocene
23.5	Miocene
33.5	Oligocene
55.5	Eocene
65.5	Paleocene

ing of geological time periods, such as the Jurassic, and the recognition that the Jurassic occurred earlier than, say, the Cretaceous. During the past century, technological developments have enabled scientists to date fossils and the deposits in which they lie with increasing precision and accuracy. Thus they are able to assign an absolute age in years rather than speak of one fossil simply being older or younger than another based on its position in the deposits. Nevertheless, the names of the time periods, coined in the eighteenth century, are still employed as a form of geological shorthand when referring to the ages of deposits and their constituent fossils. Table 1.1 lists the epochs, to give them their formal name, relevant to our discussion.

We are fortunate that the geological history of Africa has witnessed much volcanic activity because material from the volcanoes can be dated by a variety of techniques. Many of them involve measurements of the decay of radioactive elements in the ejected material, using the known half-life of the elements (the time over which the amount of the original element decays to one-half and then one-quarter and then one-eighth, and so on) to calculate the time since the formation of the sediments. Other methods make use of the periodic reversals in polarity of Earth's magnetic field, with the north magnetic pole becoming the south and the south, the north. The direction of polarity at the time of deposit formation can be preserved in volcanic materials heated above a certain temperature as well as in some sediments, because the magnetic elements line up according to the prevailing magnetic pole. A sequence of magnetic-reversal events of known age is now well established, and deposits that preserve a magnetic signature often can be matched against it.

The presence of such dateable elements in fossil-rich deposits is of immense importance for understanding the timetable of evolution in Africa. In many parts of eastern Africa in particular, volcanic marker horizons are used to provide age brackets for fossil-rich deposits termed members, and the volcanic marker at the base of a member is then named and the name is applied to the member. These markers are usually tuffs, consolidated ash ejected from a volcano during explosive eruption and often water-laid. The beauty of having such material is that it may be deposited over a vast area, especially if wind conditions are right, and the ejecta from any given eruption usually has a chemical signature that allows it to be recognized and mapped across the landscape. Since it can be dated, it is an excellent tool for stratigraphy and correlation of deposits. Moreover, such volcanic ash has even been detected in cores of deep-sea sediments, allowing the accurate dating of and correlation between terrestrial events and patterns of global change recorded in the ocean basin deposits.

Continental Drift

Africa today is joined to Europe and Asia by the Arabian Peninsula. Indeed, the African plate includes the southern part of the Arabian Peninsula, and the two finally docked with the Eurasian plate in the early Miocene, around 23.5 to 18 Ma (figure 1.1). At what point movement across the shortening gap between Africa and Eurasia became possible for terrestrial animals is less clear, although it must have been quite early because groups like the Primates are known to have migrated into Africa beginning around 33 Ma. Of course, the precise changes in the geography of the Mediterranean region before and during collision are likely to have been quite complex, but until dispersions were possible Africa had been a large, isolated island continent like present-day Australia. Indeed, Africa had been joined to Australia—along with South America, India, and Antarctica—as part of a great southern landmass known as Gondwana. That supercontinent was separated from the northern landmass of Laurasia, to which North America, Europe, and Asia belonged. By 65 Ma, the end of the age of the dinosaurs, these landmasses had long begun to break up, and the individual continents that we know today had started to move to their present positions. The total northward movement of Africa from the time of the breakup is estimated at around 14 degrees of latitude.

With the continents traveled the early mammals, at the time beginning to undergo their massive adaptive radiation. Of course, migrations had begun before the breakup of Gondwana and Laurasia, resulting in the wide distribution of many mammalian groups, such as the marsupials, which spread from what became North America over to proto-Europe and even proto-Africa as well as down through the areas that would

FIGURE 1.1

The African and Eurasian plates

As the African plate, including the southern part of the Arabian Peninsula, moved north to eventually make contact with the Eurasian plate in the early Miocene, the western portion of the Tethys Sea became isolated, forming the proto-Mediterranean.

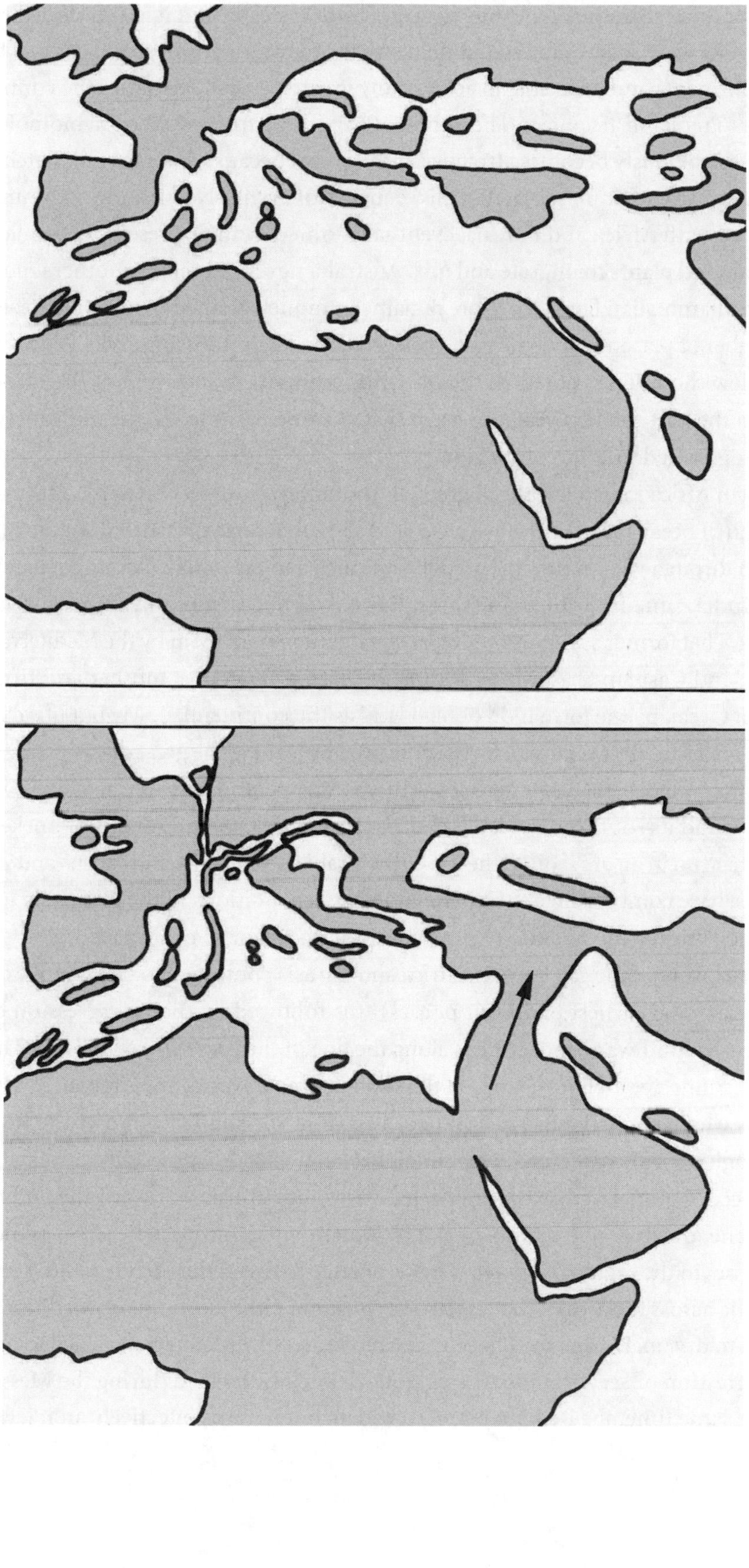

become South America and Antarctica and into Australia. But it is also clear that certain groups had their initial evolution restricted to one or another of the newly forming continents and were able to achieve any greater spread only after the continents began to reestablish contact. The notion of a unique Afrotheria seen in biomolecular analyses obviously becomes attractive at this point because it appears to match ideas about an early African isolation in this sequence of events. North and South America, together with Africa and Eurasia, eventually joined in their present form, allowing animals and plants to migrate and mix. Australia never did rejoin another landmass, and its mammalian fauna therefore remained uniquely limited to the marsupials plus a number of placental rodents that managed to make the crossing and, of course, bats that flew there. This remained the case until human colonization of the continent within the past 100,000 years, an event that at some point led to the introduction of the dingo as a domestic companion.

When Africa joined Eurasia, it gradually but intermittently closed the Tethys Sea, a portion of ocean between the two great land areas that once stretched from the Atlantic Ocean through what is now the Mediterranean Sea to the Indian Ocean. Farther to the east, India came into contact with southern Asia and assisted in the closure of the Tethys. That former seaway is now represented in remnant form by the Mediterranean, Black, and Caspian Seas, which became separated from one another and from the Indian Ocean by the formation of high land as the continents slowly but literally collided. For as the Tethys closed, so the same process formed the great east–west-trending mountain chains that run across southern Europe and into the massive Tibetan Plateau and its southern edge, the Himalayan Range. The mountains of Turkey and Iran that run in an arc around the top of the Arabian Peninsula, the Taurus and Zagros ranges, have combined with the frequently harsh conditions of the peninsula to control movements into and out of Africa over the 20 Ma or so of contact.

The closure of the gap between Africa and Eurasia continued to be a complex series of events. The earliest phase (or phases) was followed by the transgression of the Tethys Sea southward, more or less along the line of the present-day Red Sea. This left Africa connected to Eurasia only at the southernmost part of the Arabian Peninsula, across what is now the Bab el Mandeb, a strait at the southern end of the Red Sea, although intermittent contact remained between what are now southern Spain and Morocco. Contact across the Suez region was reestablished toward the end of the Miocene, during what is known as the Messinian salinity crisis, when geological processes in the region restricted water exchange between the Mediterranean and the Atlantic across the area of the present-day Strait of Gibraltar, and the Mediterranean began to dry up. During the Pliocene, the refilling of the Mediterranean led to a major transgression of seawater into the canyon of the Nile, formed during the Messinian. At the same time, the Red Sea began to widen and became effectively an inlet of the

Indian Ocean as the Arabian plate swung away, producing a breakup of the Bab el Mandeb bridge. It is clear that a semicomplete land bridge existed across the Strait of Gibraltar at some point during the Messinian—otherwise, the Mediterranean could scarcely have dried up—but there is no compelling evidence that the bridge continued to exist once the Mediterranean began to refill.

One other point to mention in connection with movements of Africa concerns the large island of Madagascar, off the eastern coast of the continent. Studies of the geology of the region indicate that the two have been separated to the extent seen today—some 400 km—for 65 Ma, since the Cretaceous, with Madagascar beginning a southward movement away from Africa as long ago as the Jurassic.

Climate Change

The climate of Earth today is very different from that of the past. Over the very long term, temperatures have fallen, although the decrease has by no means been a smooth curve. During the time of the dinosaurs, before 65 Ma, the planet was generally warmer and wetter than it is today, and temperatures were more evenly distributed so the differences between the tropics and the polar regions were less marked than they are now. Trees once grew within the Arctic Circle, and ice sheets and deserts were rare or absent. Sea levels were up to 300 m higher than at present, and thus large areas of the modern continents were covered in relatively shallow waters.

Since then, the planet has become markedly cooler, with more deserts, greater annual extremes of rainfall and temperature, and widespread sea ice and large ice sheets on Greenland and Antarctica. The Antarctic ice sheet is the largest on Earth and appears to have formed around 30 Ma, probably once the southern continents had separated sufficiently from one another to permit the circum-Antarctic current to form and trap cold waters close to the developing ice sheet. The growth of the Antarctic ice sheet was an important event in the pattern of cooling that has continued over the past 30 Ma, but the overall trend to lower temperatures has been marked by several stages of more rapid cooling. One of the most significant took place in the second half of the Pliocene, between 3 and 2.5 Ma. At the end of that epoch, the first major Northern Hemisphere glaciation is recorded from ice-rafting of debris in the northern oceans, material carried away from land by the calving of icebergs at the tips of glaciers and then dropped to the seabed as the ice melted. This global event marked the start of a long series of ice ages. The second major event is recorded close to 900,000 years ago, with a further intensification of Northern Hemisphere glaciation, so the swing between interglacial and glacial conditions became more marked and Earth began to enter a major glacial phase roughly every 100,000 years. The last of

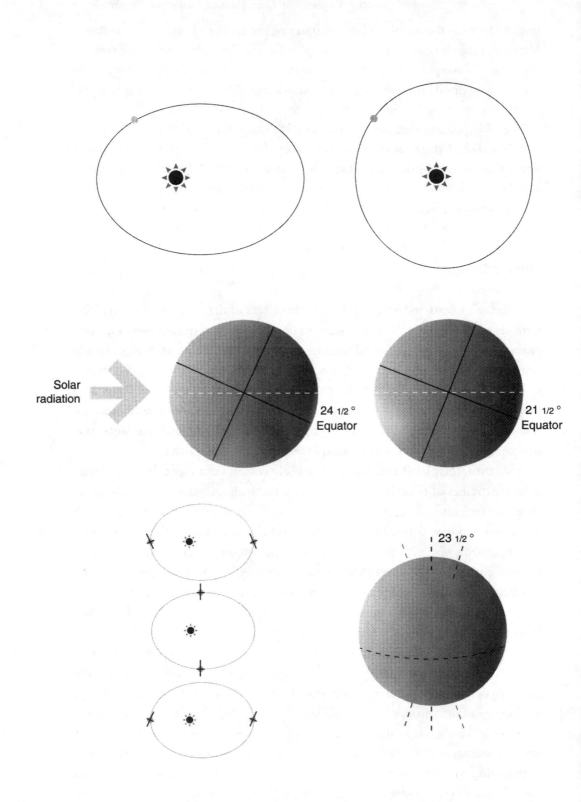

Solar
radiation

24 1/2 °
Equator

21 1/2 °
Equator

23 1/2 °

these ice ages peaked as recently as 20,000 years ago, a time when enormous ice sheets covered much of North America and northern Europe, and the seas, with so much water locked up as ice, were up to 130 m lower than at present.

Two main factors may be identified as the driving forces behind such global changes in climate. The first, of course, is the very movement of the continents, in which the breakup of Gondwana into Africa and the other landmasses of the Southern Hemisphere played a major part. Such continental drift alters the relative distribution of sea and land and changes the pattern of ocean currents, such as the circum-Antarctic current; the inevitable result is a massive change in climate, since both are major climatic determinants. In addition, the precise positioning of the continents may affect climate because land absorbs solar radiation less efficiently than water. If continental landmasses are in low, equatorial latitudes, less heat is absorbed into the global system than if open seaways are present. The tectonic events that accompany continental movements—such as rifting, volcanism, and orogeny (mountain formation)—also are important. But superimposed on all these factors are cyclical changes in the orbit of Earth itself, which determine the amount of solar energy that the planet receives (figure 1.2). We see a small-scale version of these effects in the seasons of the year as we revolve around the sun, but the longer timescale changes are even greater. When continents drift and ocean currents change, such fluctuations in solar energy are magnified; ice can grow at high altitude and high latitude, and Earth itself may then swing into and out of full ice ages.

Variations in Earth's orbit have been recognized for many years and are the basis of an important theory of the cause of ice ages put forward by the Yugoslavian physicist Milutin Milankovitch in the 1930s and refined since then to provide a clear and consistent explanation. There are three major components of orbitally induced fluctuations in solar energy. The first is the shape of the orbit, or eccentricity, which varies from more circular to more elliptical and then back over a 100,000-year period. A more elliptical orbit produces greater contrast in the seasons in either the Northern or the Southern Hemisphere, depending on whether Earth is nearer to or farther from

---

FIGURE 1.2

Orbital changes and climate forcing

The variations in the orbit of Earth interact with one another to alter the amount of solar energy received over time. (*Top*) Eccentricity of the orbit. As the shape of the orbit varies, the seasonal contrast is enhanced or reduced. (*Center*) Tilt of the axis of rotation of Earth in relation to the plane of the orbit. This tilt produces the seasonal contrasts as each hemisphere points toward (summer) or away from (winter) the sun. As the tilt reduces, the polar regions receive less sunlight in summer. (*Bottom right*) Precession or wobble of the axis. This wobble helps produce the "precession of the equinoxes" (*bottom left*), the changes in the coincidence of summer and winter with distance from the sun that results from the shape of the orbit. Thus if the Northern Hemisphere is tilted toward the sun (summer) when Earth is farthest from the sun, the summers will be relatively cool.

the sun. The second component is the variation in the tilt of the rotational axis of Earth in relation to the plane of the orbit around the sun, which changes from 21.5 to 24.5 degrees and back over a 41,000-year cycle. Greater tilt results in hotter summers in one hemisphere and colder winters in the other, as either the Northern or the Southern Hemisphere leans toward or away from the sun. The third component is the wobble of the tilted spin axis, or precession, over a cycle of 23,000 years, which alters the time in the orbit at which the seasons fall.

Because the variations in solar energy input and their combined effect are cyclical, the pattern of climate change that they induce is also cyclical. Earth has experienced around eight major ice ages over the past 800,000 years, with the pattern repeated, perhaps with less intensity and a different frequency, back to 2.5 Ma and the time of the onset of the first major Northern Hemisphere glaciation. How do climatologists know all this in such detail? The orbital variations, continental movements, and seaway closures are, of course, established facts of astronomy and geology, but the climatic effects of these processes, essentially the output of the system, are recorded in geological deposits. Now that we can recognize both the input to and the output from the system, we can understand the relationships between the two in increasing detail. Ice sheets produce typical patterns in the landscape, while sediments that accumulate leave clues about aridity and temperature of the time of formation, and biological remains in the sediments add further information about local or regional conditions.

Unfortunately, however, large ice sheets have tremendous erosive power, and the very processes in which we are interested frequently destroy the signs of earlier events in the sequence. The one place on Earth where the evidence of past global climates accumulates and tends to be preserved over a reasonably long span of time is the sediments of the deep ocean. Such sediments are relatively unaffected by runoff of material from the continental landmasses and instead consist largely of the remains of sea creatures, particularly microscopic, single-celled shelled organisms known as foraminifera. The calcium carbonate of the shells of foraminifera is derived from seawater, and the chemical composition of the shells reflects the composition of the seawater. Oxygen 16 (^{16}O), a relatively light isotope, is taken up preferentially when seawater evaporates, leaving behind the heavier isotope ^{18}O. In interglacial conditions, like those of the present day, the ^{16}O is returned to the oceans by runoff waters from the land, but during glaciations it remains locked up in the ice sheets. During ice advances, sea creatures therefore construct their shells from seawater with an increased ratio of ^{18}O to ^{16}O, and the sediments that they form on death reflect that heightened ratio. A core taken from the ocean sediments shows a fluctuating ratio of the two isotopes, and thus directly reflects the changing pattern of global cooling and warming. The complex sequence of changes in such cores precisely matches the orbitally induced variations in solar energy received by Earth over the same period.

But, of course, all of this raises the question of why the glacial–interglacial sequence began near the end of the Pliocene. If the Antarctic ice sheet had been in existence for millions of years and the orbital variations for many times that period, what could have triggered such massive climate change? To understand such events, we probably have to go back to the question of continental movements and ocean currents. One view, put forward particularly by the paleontologist Steven Stanley, suggests that the narrow neck of land between North and South America, the Isthmus of Panama, played an important role.

The North Atlantic and the landmasses that border it are heated in large part by warm and salt-enriched water transported northward by a vast conveyor belt of shallow current that arises in equatorial regions of the Pacific, travels around Africa, and continues across to the Americas. Here it is driven northward, reinforced in its movement up the Atlantic by the wind-driven Gulf Stream. At the northern extremity, somewhere between Greenland and Norway, the warm water gives up its heat to the atmosphere and cools sufficiently to increase in density and sink, traveling southward at depth back to the Pacific, where it is warmed and rises again to drive the conveyer. The North Atlantic water of the system sinks before it can enter the Arctic Ocean, which thus remains cold.

If there were no Isthmus of Panama between North and South America, Stanley has argued, the conveyor system would disappear, since the currents would carry the warm water around the top of an island-like South America. The northern latitudes would be deprived of the heat transported by the conveyor belt. But without the warm and dense salty water to sink so rapidly, it is likely that the smaller amount of heat carried by a reduced Gulf Stream would be able to travel farther north and induce some warming of the Arctic Ocean. A warmer Arctic would be ice free and, in turn, would not send cold air southward to cool the northern latitudes. That may have been the situation until the mid Pliocene, at about 3.5 Ma, because the Isthmus of Panama came into being only between 3.5 and 3 Ma when North and South America finally came into contact, pushing up land in the join. With that closure of the gap between the continents, the conveyor was switched on and the orbital variations, waiting so to speak in the wings, began to exert control, swinging Earth into what by 2.5 Ma was a series of major ice ages. In this scheme, we may therefore see events in Central America and their effect on the Arctic and the development of fully glacial conditions as a continuation of processes set in train by the earlier formation of the Antarctic ice sheet.

As with many things in science, however, there are alternative interpretations. Other researchers argue that the conveyor system did not increase in strength in that way at that time, while a more recent proposal suggests that continental movements in the Pacific were of greater importance. Geologists Mark Cane and Peter Molnar have highlighted the effects of the northward drift of Australia and New Guinea over the past 5 Ma. According to their model, the movement of the islands would have narrowed the seaway that allowed warm water from the South Pacific into the Indian Ocean, and replaced the warm water with colder water from the North Pacific. The immediate impact would have been a reduction in evaporation and hence precipitation in eastern Africa, although Cane and Molnar also argue that the narrowing of the seaway would have slowed the transportation of heat to higher latitudes and, in the process, aided the growth of ice sheets in the Northern Hemisphere. Whether one or another view is correct, or some subtle combination of such processes occurred in both the Pacific and the Atlantic, only time will tell, but it is clear that climate change and continental drift have taken place and that the link between them is likely to be close.

Evolution

We learn more and more each year about biology and the genetic basis of evolution and can make strong inferences about the relationships between living species based on their genetic makeup. But the fossil record is *the* record of life on Earth. However much we come to understand the genetic basis of life as a result of our work in the field and laboratory, it is to the fossil record that we must turn if we wish to determine the sequence of events that led to the present-day diversity of living things. Without that record, we would never know that dinosaurs, mammoths, saber-toothed cats, or even our own extinct relatives had existed.

The sheer variety in size and morphology of past life-forms is astonishing, but perhaps the outstanding feature of the fossil record is change—at times rapid, at times slow, but always evident. Such evidence of the impermanence of any species, however well adapted to its environment it seems to have been, leads to a question: What is the motor of evolution? This question has exercised biologists ever since Charles Darwin expressed his ideas about natural selection in 1859 in one of the most important books ever written, *On the Origin of Species by Means of Natural Selection*. Various more or less fanciful answers have been put forward, from simple assumptions that new species just come into being after a certain amount of time has passed, to more complex models such as the Red Queen, according to which organisms, like the character in *Through the Looking-Glass*, are perpetually running hard to stay in place, driven by the motor of endless competition with others in their environment. In our view, no satisfactory answer to the supplementary question of how either accumulated change or constant competition actually causes speciation has been forthcoming from these suggestions. A more likely answer comes when we consider both the nature of species and the mechanism of change.

In sexually reproducing species like ourselves and other mammals, the offspring receive genetic information from both parents, information that each parent, in turn, received from its parents. Thus it is necessary that males and females recognize each other as potential mates, and the species may therefore be regarded as simply a group of organisms bound together by a shared system of mate recognition and fertilization in what the zoologist Hugh Paterson has termed the Recognition Concept of Species. The shared system involves a signal-and-response chain between potential mates—the precise form of which varies but may include calls, chemical signals, color and patterning of pelage or plumage, and movements and display in often complex courtship rituals—and depends on the habitat in which the species lives.

Natural selection is intensely conservative. Although it may fine-tune an organism to its environment, natural selection also operates against extremes in characters as well as components of the fertilization system to ensure that they remain stable while the organism stays in its normal habitat. Thus a species persists if conditions remain constant and the system remains stable. If conditions change, however, the population may find that the new circumstances affect its lifestyle, in which case it must adapt or move to a more suitable area if it is not to become extinct. If the new conditions fragment the habitat, isolate a small part of the original population, and affect the fertilization system, natural selection may operate relatively rapidly to change the components of the system in the subpopulation. Mating between members of the new and the original, parent population may become impossible because their fertilization systems differ from each other, and so the subpopulation has evolved into a new species.

The Recognition Concept of Species thus elucidates two major characteristics of the evolution of life seen in the fossil record: the development of features that fit organisms to their environment and the appearance of new species. It also explains the relationship between speciation, extinction, and dispersion. A new species that originates from a subpopulation of the parent species necessarily has a localized place of origin, and distribution beyond the range of the subpopulation must therefore be achieved by dispersal. Species with narrow specializations are relatively constrained in their range of territory, while environmental generalists, such as large carnivores, are much freer to move. Specialists, even those that remain in the same place, are likely to encounter new habitats more frequently than generalists during times of environmental change. Thus they tend to face conditions that induce speciation (or extinction) more often than generalists, leading to differing rates of evolutionary change in lineages. In the absence of physical barriers, habitat dictates where given species can and cannot exist, and habitat is largely determined by a combination of the prevailing climatic conditions and the physical environment.

When species and the process of speciation are considered in this way, it is evident that most of the major alterations in the development of life—those involving dispersion, extinction, and speciation—are likely to have been triggered by changes in the physical environment and their effect on habitats, as the paleontologist Elisabeth Vrba has pointed out. In the absence of such changes in geography and climate—which can be considered the ultimate motor of evolution—species, particularly those that can tolerate a wide range of conditions, tend to persist, which is why we see variety in the patterns of evolution in the fossil record. In addition, and perhaps most important in the context of the evolution of the mammalian fauna of a continent, such an external motor is likely to cause major changes across a wide diversity of lineages to occur in concert—what Vrba has termed an evolutionary "turnover pulse," evidence of which is apparent in the overall patterns in the fossil record of Africa.

2 THE BACKGROUND TO MAMMALIAN EVOLUTION IN AFRICA

THE BACKGROUND TO THE EVOLUTION of mammals in Africa—the physical geography and biotic environments of the continent—is the result of continental drift and other geological forces, global climate change and local habitat conditions, and floral and faunal distributions that reflect these tectonic and climatic trends.

Physical Geography

Africa is large. At around 30 million square km, it forms about one-fifth of the land surface of Earth, and with almost one-half of its land at over 1000 m and most of the area south of the Sahara above 500 m, it is surprisingly elevated in comparison with other continents (figure 2.1). The basins of the Kalahari and the Congo are simply regions of lesser elevation rather than depressions, although there are areas in both Egypt and the Afar Depression of Ethiopia where the land surface is several tens of meters below sea level.

Based on its physical geography, Africa can be divided into several component parts. Low Africa, with an altitude largely below 900 m (although some parts of the central Saharan region dome above this), is that portion south of the Atlas Mountains, a range in the northwestern corner of the continent, and generally north of a line running eastward from the southwestern margin of the Congo Basin and then northward along the western edge of the eastern African highlands to the Red Sea coastline just below Suez. High Africa, to the south and east of this line and gen-

FIGURE 2.1

The topography and major watercourses of Africa

erally over 900 m, consists of the eastern African highlands and the elevated southern platform.

Much of Low Africa consists of the Sahara (figure 2.2). Stretching across 2000 km of northern Africa, it is the largest desert on Earth and almost one-quarter of the total land area of the continent. It does not consist simply of sand; although *ergs*, as the sand dunes are known, are extensive, they make up only around 10 to 15 percent of its area. The biggest, the Great Eastern Erg of Algeria, is the size of a larger western European country, such as France or Germany. The Sahara also includes *regs*— hardened areas of stone, gravel, and silts—and substantial rocky plateaus known as *hamadas*, which reach to more than 3000 m. The desert is bounded to the northwest by the Atlas Mountains (figure 2.3), which rise in places to 4000 m and whose northern outpost, the Rif Mountains of Morocco, are easily visible from far inland in the mountains of southern Spain. In High Africa, the equatorial highlands that surround the Great Rift Valley of eastern Africa reach to over 5000 m, while the grassland savannas of eastern Africa may be up to 1500 m above sea level on the East African Plateau, as high as the highest point in the entire British Isles, to give one simple comparison. The Rift Valley itself lies at altitudes of up to 2000 m in places. The wide landscapes of the South African highveld are at 1000 m and more, and the mountains of Lesotho rise to over 3000 m. Figure 2.4 shows the impressive Drakensberg escarpment, which forms the eastern boundary of the highveld. The Great Karroo, an arid inland plateau that makes up much of the Cape Province of South Africa, reaches elevations between 1000 and 1500 m. In southwestern Africa, the north–south-trending highlands of Namibia, which parallel the coast some 100 km inland, rise to more than 2500 m, while in the Namib Desert, which lies between the highlands and the Atlantic Ocean, some of the world's largest sand dunes are more than 300 m tall.

Areas of the interior plateaus are remnants of ancient land surfaces, part of the terrain of the supercontinent of Gondwana, which incorporated Africa, India, South America, and Antarctica. The mountains, in contrast, particularly those of eastern Africa, are generally much younger, the result of a combination of uplift, volcanism, and tilting that has gone on during the past few tens of millions of years. As a result, the transitions from plateau to mountain are often rapid, producing enormous escarpments such as the east- and south-facing 1000 m rampart of the Drakensberg of South Africa. Only those mountains of the northern Atlas and southern Cape ranges show evidence of the folding seen so commonly in the mountains of other continents.

Evidence of terraces and shorelines, along with alluvial deposits and deltas, suggests that many parts of Low Africa were covered by considerable bodies of water during the Miocene and Pliocene, with large lakes in the Congo Basin, the western Sahara, and the Sudd area of the Upper Nile and a greatly expanded Lake Chad to the

FIGURE 2.2

Two aspects of the Sahara Desert

(*Top*) Sand dunes in the central Sahara, Chad. (*Bottom*) Sand dunes, rocks, and distant water pools in the western Sahara, Mauritania. (Photographs courtesy of Pablo Peláez [*top*] and Clare Milsom [*bottom*])

FIGURE 2.3
The Atlas Mountains
The snow-capped Atlas Mountains of northwestern Africa, as seen from a fluvial valley. (Photograph courtesy of Susana Fraile)

south of the Sahara. Today large parts of these areas are marshy terrain, and the ancient water bodies are likely to have existed in part because the internal drainage basins lacked outlets to the sea, although the Congo Basin has now achieved that with the Congo River cutting through the highlands to the Atlantic coast. The flatness of the Congo Basin is underlined by one telling statistic: in the 2000 km or so of the great northern loop that the river makes between Kisangani and Kinshasa, it falls less than 100 m and moves slowly in channels that may be up to 15 km wide. Despite the small drop, such is the size of the Congo River and of its catchment area that the discharge is second only to that of the Amazon. Whether the growth of these former bodies of water was also due to an increase in precipitation, or to precipitation in conjunction with lowered temperatures (and hence evaporation), is less clear. In High Africa, the present-day Okavango Delta of northwestern Botswana (figure 2.5), also may have been represented by a much larger lake.

The Great Rift Valley of eastern Africa is one of the most important single features of the continent's landscape (figure 2.6). It is a massive fracture that runs southward for more than 6000 km from the Jordan Valley to where the Zambesi River runs into the Mozambique Channel, between the east coast of Africa and Madagascar. The scar produced by this fracturing is actually visible from space. The Red Sea owes its origin

FIGURE 2.4
The Drakensberg escarpment
The view eastward from the top of the northern end of the Drakensberg escarpment, to the east of
Pretoria, South Africa. (Photograph by Alan Turner)

to the rifting process; Africa and the southern part of the Arabian Peninsula are, after
all, a single plate, and the movement that has produced the Red Sea is a secondary
event that began after the African and Arabian plates docked with the Eurasian plate.
The western arm of the Rift Valley, which extends to Mozambique, has produced the
lake chain from Lake Albert to Lake Tanganyika and Lake Malawi, some of the deep-
est lakes in the world. The large but quite shallow expanse of Lake Victoria, which lies
between the arms with a maximum depth of around 80 m, is a relatively recent phenomenon produced by naturally blocked drainage. Along the eastern arm of the Rift
Valley lie several smaller lakes, of which Lake Turkana, near the junction of the western and eastern arms, is the largest.

The rifting of eastern Africa has occurred in conjunction with phases of uplifting
and tilting of the continent as a whole since its separation from Gondwana. Uplifting
and rifting are closely interlinked, and both are related to the volcanic activity that
has been a prominent feature of the eastern African landscape and environment
down to the present day. In the mid-1960s, Ol Doinyo Lengai, on the eastern side of

FIGURE 2.5

The Okavango Delta

An aerial view of the distal part of the Okavango Delta of northwestern Botswana during the dry season. An elephant provides scale against which to measure the size of a termite mound, which are essential for the growth of islands of firm ground in many swamp areas of Africa. (Photograph by Mauricio Antón)

FIGURE 2.6
The Great Rift Valley
The bottom of the Great Rift Valley in Tanzania, with Lake Manyara in the distance and the Great Rift escarpment to the right. (Photograph courtesy of Luis Luque)

the Serengeti Plain in Tanzania, just to the south of Lake Natron and about 180 km to the west of Mount Kilimanjaro, erupted quite spectacularly and covered its cone with ash. This was the fifteenth or so eruption in about 100 years, and the ash from these and other eruptions has mainly been carried westward by the winds and has had a marked effect on the growth of grasslands in the Serengeti. The eastern arm of the Great Rift Valley still has around 30 active volcanoes, and many of the lakes in this part of the system show evidence of geothermal activity in the form of hot, mineral-rich waters and their resultant, often extensive, deposits of salts (figure 2.7). Indeed, only two of the dozen or so lakes in the eastern arm, Baringo and Naivasha in southern Kenya, are truly freshwater bodies, while Magadi and Natron, just farther to the south on the Kenya–Tanzania border, are highly alkaline to an extent not found even in the Danakil Depression of Ethiopia.*

Thus until around 30 Ma, just within the period of major concern in this volume, the topography of eastern Africa was rather different from that of the present day At around that time, there is evidence of extensive, even catastrophic, volcanic activity in the Afar region of Ethiopia close to the Red Sea. The Great Rift Valley appears to have begun to form, to become an established feature of the landscape in northern

---

*Not surprisingly, *magadi* means "soda lake" in the local Masai language, while "natron" is the word for naturally occurring sodium carbonate, derived by way of the Arabic *natrun* (niter) from the Greek *nitron*.

FIGURE 2.7
Lake Bogoria

A large group of lesser flamingos clusters around an active geyser in Lake Bogoria, in the Great Rift Valley of Kenya. In the far background, the escarpment of the Great Rift dominates the horizon. (Photograph by Mauricio Antón)

Ethiopia, by the mid Miocene, but major changes to the landscape were still occurring. It is estimated, for example, that parts of the Afar Plateau in northern Ethiopia have been elevated by close to 2000 m within the past 8 Ma. Averaged to 0.25 mm per year, or 1 m every 4000 years, that may not seem very much as a day-to-day phenomenon, but the total effect on the topography over that time is immense. Moreover, the average figure is misleading, since most of the change may have been effected over a much shorter time span between perhaps 3.5 and 2.5 Ma, with the Afar Plateau uplifted by around 1000 m while the Rift Valley to its south subsided by up to 2000 m. Such changes would have had noticeable effects within the span of a few generations of longer-lived mammals.

The Danakil Depression, a triangular area at the northernmost end of the Great Rift Valley on the Red Sea coast of Ethiopia, provides perhaps even more impressive evidence of geological forces at work. Behind a ridge of higher land now separating it from the sea, parts of the depression lie 130 m below sea level, having sunk as the Rift Valley has opened. Several active and dormant volcanoes cluster there, and boiling-hot water laden with salts emerges from springs. The Awash River, along which several important sites containing fossils of our own relatives and evidence of their technological development have been found, emerges from the Ethiopian highlands, only to disappear into Lake Abbe, from which there is no outlet. Evaporation in the furnace-like conditions of the Danakil takes care of the water.

The flanks of the eastern arm of the Great Rift Valley, in particular, owe much of their total elevation to volcanic growth from around 15 Ma onward, with major volcanoes added during the past 5 Ma, as well as to substantial uplift in the past 3 Ma. The western arm of the Rift Valley appears to have begun to form somewhat later, around 12 Ma, with subsequent faulting after 5 Ma. The rift flanks exceeded 4000 m in the Ruwenzori Mountains of western Uganda, which today peak at 5109 m, and much of this uplift appears to have occurred between 3 and 2 Ma. Nor should we forget that the present-day topography so influenced by tectonic and volcanic activity is merely the current stage; earlier volcanoes have eroded away and formed deep yet relatively young deposits from their sediments, and the process of change will continue.

By the end of the Miocene, the topographic changes in eastern Africa were more than sufficient to start to have a major impact on the climate of the region. Temperatures decrease with altitude, and mountains cause air to rise and shed moisture, producing conditions suitable for a rain forest on one side and a rain shadow on the other. Such effects may be highly localized, producing a mosaic of environments and habitats in a region where more regular terrain would result in more uniform conditions.

The folded mountains of northern Africa appear to be of late Oligocene age and are clearly a product of the contact between the African and Eurasian plates. The folding lifted much of the northern region above the shoreline of the Tethys Sea, although substantial areas to the east of the Atlas Mountains, in what are now Libya and Egypt, were covered once more by shallow seas during the Miocene. In southern Africa, the uplift that produced, among other features, the highveld and the Drakensberg Escarpment, was similarly large scale. There appears to have been a substantial uplift of the southeastern hinterland during the Pliocene, perhaps by as much as 1000 m, superimposed on a rise of around 250 m that had occurred during the Miocene. The absence of rifting in southern Africa is a notable difference with eastern Africa, although subsidence during the Pliocene may have occurred in the Bushveld Basin in the north of the Transvaal, with resultant changes in the environments of the fossil-rich localities there.

Climates

Africa sits astride the equator—indeed, it is the only continent with substantial land-masses on either side of it—and much of it falls within the tropics. Thus climates generally range from warm to hot—in some places, such as the Danakil Depression, very hot—with extremes of cold limited to highland and mountains. The continent as a whole is relatively dry, so the distribution of vegetation is essentially limited by rainfall or, to be more precise, by the balance between water demands and water availability. Figure 2.8 shows the general pattern of rainfall across Africa, which is related to its equatorial position, with the northern and southern halves of the continent being generally drier during their respective winters. The exceptions are the Mediterranean areas of the northern coast and the Cape of the southern, where there is winter rainfall, and the equatorial regions, where there are two wet seasons because the seasonal changes of the two hemispheres interact. Because of the seasonal nature of rainfall, the rivers of Africa vary greatly in their water volume throughout the year; the annual flooding of the Nile and the renewed fertility brought by the silts was, of course, a major factor in the success of Egypt during the pharaonic period. The large amount of rain that can fall in the wet season is the reason for the size of such swamplands as Okavango in Botswana, Lake Bangweulu in northern Zambia, and Lake Chad on the Nigeria, Cameroon, and Chad border, where rivers often have no other drainage.

The physical geography of Africa—with the northwestern, eastern, and southern coastal areas separated from the interior by mountains and escarpments—also plays a great part in the distribution of moisture when it is available. Thus, to take one example, the Mediterranean floras of coastal Morocco and the South African Cape do not penetrate far inland, and semiarid and fully desert areas with increasingly sparse vegetational cover are soon encountered. The contrast between coast and interior is seen particularly well in Algeria, the second largest country in Africa, which is very long from north to south. Most parts of the Sahara Desert receive less than 100 mm of rain per year, although moisture may be obtained from dew formed during the cold nights—daily fluctuations in temperature of 56°C down to −10°C have been recorded in southern Algeria. Moisture from a similar source plays an important part in the ecology of the Namib Desert of southwestern Africa, where the cold water of the Atlantic Ocean that results from the Antarctic Benguela current produces regular sea fogs that may blanket the sands for up to 30 km inland. An increase in rainfall on the edges of the fully desert area is enough to produce a change back through grassland to lightly covered parkland.

Annual rainfall in Africa is distributed in a broadly horseshoe-like pattern around

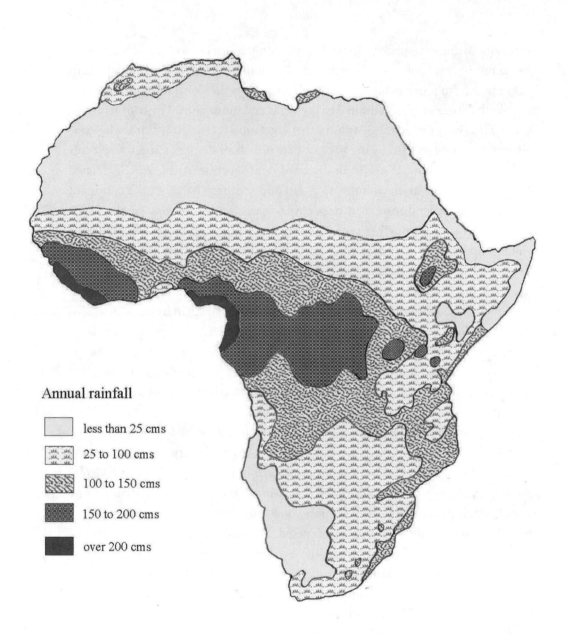

Annual rainfall

less than 25 cms

25 to 100 cms

100 to 150 cms

150 to 200 cms

over 200 cms

FIGURE 2.8
The annual distribution of rainfall in Africa

the equator (Figure 2.8), tailing off and becoming more patchy to the east, where the highlands induce rain shadows, diminishing the fall of moisture brought in off the Indian Ocean. The reason for the precise rainfall distribution is complex and tied up with the global wind patterns, the topography of Africa, and even the effects of the large inland lakes, such as Lake Victoria and the lakes of the western Great Rift Valley. But in essence, the pattern results from the equatorial areas having two wet seasons. High-pressure zones outside the tropics produce winds that move northward (in the Southern Hemisphere) and southward (in the Northern Hemisphere) toward the equator and the zone of lower pressure produced when heated air rises. The rotation of Earth causes the winds in the Northern Hemisphere to be deflected clockwise and those in the Southern Hemisphere to be deflected counterclockwise, a phenomenon known as the Coriolis effect. This produces what is termed the Intertropical Convergence Zone (ITCZ), where winds from north and south meet and, by turning westward, reinforce each other in bringing moisture from the Indian Ocean over eastern Africa. The high-pressure zones shift northward and southward with the seasons, and thus the ITCZ moves from being roughly at the equator in January (the Southern Hemisphere summer) to perhaps 20° N latitude in July (the Northern Hemisphere summer). When it is to the south, at the equator, the ITCZ brings winds mainly from the east, off the Indian Ocean, although western Africa south of the equator continues to receive winds off the Atlantic Ocean. When the ITCZ is north of the equator, the northward-moving winds are now traveling in the Northern Hemisphere and thus bend to the east, carrying moisture from the Atlantic deep into central Africa, although the highlands produced by the rifting process prevents this moisture from reaching the eastern side of the continent. In comparison with equatorial Africa, northern and southern Africa receive relatively much less moisture, particularly the north and especially the northeast, where most of the winds are traveling southward or southwestward throughout the year and do not pass over bodies of water from which to pick up moisture.

When we turn to the past, and to the question of climate change over time, various problems face us. One is to match the larger-scale patterns of global climate change reflected in the cores recovered from deep-ocean sediments with events in the less continuous and complete terrestrial record. In northern latitudes, where large ice sheets were particularly active, such matches have been achieved to a certain extent, especially for the last few glacial–interglacial cycles. In Africa, however, such terrestrial information that can be so closely linked to these cycles, and thence to the ocean cores, is generally not available.

In broad terms, it is likely that each of the major ice ages produced lower temperatures and decreased precipitation in Africa. The precise effect of this change in climate on the growing conditions for plants is then a matter of some conjecture. If decreased precipitation led to increased aridity, the dry conditions are likely to have fragmented the rain forests across the tropical region, while allowing the drier areas to the north and south of the continent to expand toward the equator. Fossil sand dunes show that the Sahara extended 450 km farther south around 20,000 years ago, at the height of the last glaciation. Each return to interglacial conditions would have reversed the process, rejoining the areas of fragmented rain forest and driving the arid regions once more to the north and south. The present interglacial began around 10,000 years ago, and at around 6000 years ago reached a peak during which increased moisture seems to have made currently barren areas of the Sahara habitable by plants, animals, and humans. Intermediate conditions, the interstadials when the climate ameliorated but did not return to the fully interglacial state, would have produced similar change on the continent on a smaller scale. The climate, and with it the rainfall and thus the vegetation, effectively would have pulsed on a timescale correlated with that of the glacial cycles. The impact on the fauna is likely to have been profound. Any animal able to stand still over a long period theoretically would have witnessed such cycles in the vegetation, but, of course, nothing lives that long. Species adapted to rain forest, or even to just woodland cover, would have had to contract their ranges during glacial phases, adapt to drier conditions, or go extinct. Species adapted to arid habitats would have faced the same problems during the swing back to interglacials, as expanding forests encroached on and fragmented the arid areas. Contracted ranges in both cases would have led to population fragmentation.

Perhaps, though, the complex topography of High Africa, in particular, produced local and regional habitats that buffered the effect of climate change, reducing the extent to which evidence of high-latitude responses to shifts in Earth's orbital parameters can be applied to equatorial regions. As the paleobotanists Eileen O'Brien and Charles Peters have pointed out, decreased precipitation combined with lower temperatures also may have altered the balance between water requirement and water availability to the advantage of plants, again ameliorating the impact of climate change. Indeed, previous models of past climate changes in Africa had developed what was known as the Pluvial Theory, which suggested that glacial events in temperate regions of the world were matched by wet conditions in Africa. The idea was based on arguments about the extent to which lowered temperatures would have decreased evapotranspiration by plants, but this interpretation has generally fallen from favor. O'Brien and Peters have also stressed that the large inland bodies of water that once existed in central and northern Africa may have had a major damping effect on climatic excursions, particularly during the Pliocene.

What is recorded in Africa is a fairly good correlation with the glaciation of 2.5 Ma, however. Pollen in sediments from the Shungura Formation and from Hadar, both in Ethiopia, points to increasing cold and aridity in the eastern part of the continent around that time and to the spread of grasslands. This evidence is supported by changes in the composition of small-mammal faunas in the Shungura deposits. More directly still, deep-sea cores from the South Atlantic even preserve windblown sediments from increasingly arid terrestrial environments of western Africa during the same period. The question remains whether such indicators are explained as local responses to tectonic or volcanic events rather than global phenomena, although features of turnover in the mammalian fauna as a whole do suggest that a more continent-wide trend is involved.

Biomes

The modern African flora contains about 40,000 identified species, and the range of variation in associations—the vegetation at any one place or in any one region—is considerable. It is estimated that humans can eat around 5 percent of the modern plants, but, of course, edibility by humans is only one feature of the importance of the vegetation. Shade, shelter, refuge, medicine, and food for animals that are themselves eaten by humans are often equally valuable characteristics to us and to the animals. Moreover, the vegetation itself has important interactive effects on climate at various scales and on the physical environment. The plants themselves have, of course, evolved, just like the animals.

Efforts to understand the vegetation of a whole continent can quickly become bogged down in a mass of detail. In sub-Saharan Africa, around 190 vegetational types have been defined, each with its own flora and, of course, fauna. The Karroo and Namib region alone has been divided into no fewer than 15 types of arid savanna. For a more general discussion, a less detailed system is required, such as the major biomes with the vegetation distributed as in figure 2.9. A biome is a block of vegetation with its associated fauna and is used as a convenient shorthand by ecologists and biogeographers to map and describe the distribution of ecosystems using such terms as "tundra," "boreal forest," "savanna," and "tropical forest." In its general sense, a biome is a taxon-free designation; tropical forests exist in various parts of the world and contain different mixes of plant and animal species, but all tropical forests have similar characteristics of vegetational structure, growth, and production.

The biomes of Africa consist mainly of Mediterranean or chaparral vegetation, desert, savanna, and tropical lowland and montane rain forest. Swamplands are of considerable importance in some areas, although on a continent-wide scale they currently make up a small proportion of the total range of biomes. In very broad terms, as pointed out by the geologist Basil Cooke, the vegetation can be envisaged as a series of horseshoe-shaped belts centered on the equatorial rain forest of western central

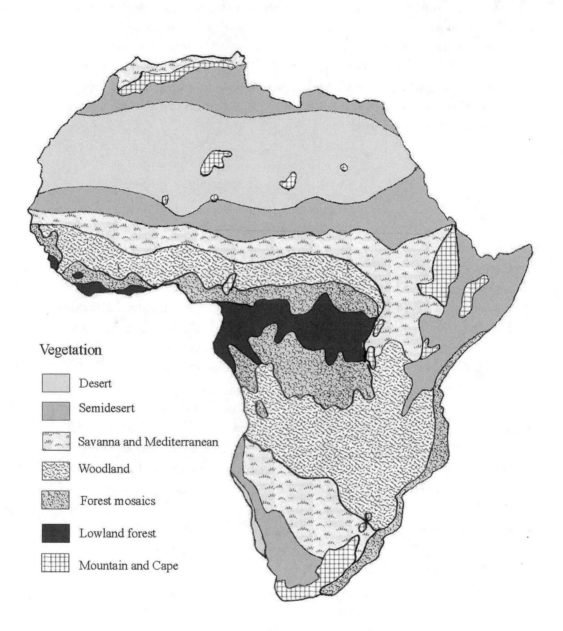

Vegetation

	Desert
	Semidesert
	Savanna and Mediterranean
	Woodland
	Forest mosaics
	Lowland forest
	Mountain and Cape

FIGURE 2.9
The distribution of vegetation in Africa

Africa. This pattern essentially mimics that of the distribution of rainfall, reinforcing the point that humidity rather than temperature is the limiting factor.

Savanna

South of the Sahara, which dominates northern Africa, savanna and open woodland extend north and south of the equator, covering about 40 percent of the continent. "Savanna" will be widely familiar as the word used to describe much of the African landscape—indeed, it is often thought of by many as the typical Africa, with extensive grasslands teeming with herbivores and their predators. There is truth in that view, especially in relation to prey and predators, but it is not the whole truth. A rich array of palatable grasses that support an equally rich variety of grazing species like zebras, white rhinos, wildebeests, and gazelles are found there on a seasonal basis, according to rainfall, together with lions, cheetahs, leopards, hyenas, and wild dogs. But the term "savanna" encompasses a number of tropical vegetational types, ranging from almost pure grassland to woodland with a high proportion of grassland (figure 2.10; plates 1–8).* The co-occurrence of grass and trees is one of the major characteristics of savanna, conditioned by the interaction of a number of factors, including moderate, seasonally distributed rainfall; warm temperatures; medium altitude; often shallow soils that inhibit tree growth; and periodic fires initiated by lightning strikes. In areas where suitable soils, altitude, and temperature combine with greater levels of rainfall, trees may establish themselves and start to predominate over grassland—the extreme is rain forest.

The ecologist Norman Owen-Smith has pointed out that the growth of pure grassland generally depends on local, very specific soil conditions, and the vegetation of the Serengeti Plain in Tanzania serves to highlight his point. In the east, under the influence of numerous eruptions of ash from Ol Doinyo Lengai, the soils prevent anything except short grasses from getting established. But in the west and north, better drainage permits longer grasses and, eventually, trees, albeit thorny and relatively short and separated from one another by grass cover—really a wooded grassland. River courses may even be bordered by taller and denser riverine forest, although seasonally inundated floodplains or other areas underlain by clay may be restricted to grass cover. Other aspects of savanna type also depend on the soil nutrient status stemming from soil structure and the persistence of moisture and nutrients. Owen-Smith places true African savannas with trees and grass in two main categories: fertile savannas with fine-leaved trees, such as acacias, and infertile savannas with broad-

---

*The word "savanna" comes from the Spanish loan word *sabana,* which was derived by the Spanish from the Taino word for woodlands that contain varying amounts of grassland.

FIGURE 2.10

Two aspects of savanna

(*Top*) A woodland savanna in the Savuti Marsh of Chobe National Park, Botswana, with a herd of impalas and a pair of elephants near a termite mound. (*Bottom*) A grassland savanna in Samburu National Park, Kenya, with an oryx in the grasses and Doum palms and riverine woodland in the distance. (Photographs by Mauricio Antón)

leaved trees. The differences in fertility may be highly correlated with herbivore biomass, since woodland in infertile savannas may provide plenty of browse, but the
leaves are often unpalatable.

Among the more wooded areas of the savanna, the most widely spread is *miombo*,
a vegetational type of the central African plateau of Tanzania and Zambia that grades
into true woodland. It is dominated by tall, leguminous trees—species that grow their
seeds in pods—such as members of the genus *Brachystegia*, instead of the acacias of
the more open bush. Wooded parts of the southern region of central Africa, especially
flatter and wider valleys, tend to be characterized more by mopane vegetation, which
derives its name from the dominant tree, *Colophospermum mopane*. The baobab
(*Adansonia digitata*), also known as the upside-down tree because of its rather root-
like branch structure, is also found in such areas. Of course, human activity impinges
on the distribution of these wooded areas, and deliberate fire-setting, sporadic cul-
tivation, and overgrazing often produce coarse grasslands.

Rain Forest

Rain forest is generally a lower-altitude biome, the product of the warmth and
humidity of western equatorial Africa, although it is also found in the eastern mountain areas where rainfall is sufficient to produce and maintain year-round growth
even at slightly higher altitudes (figure 2.11). In the western lowlands, in particular,
it is characterized by extensive forests of tall trees whose huge spreading crowns produce a dense canopy at a height of 30 m or so above the ground, beneath which
smaller trees and vines may form lower canopies. Vegetational productivity may be
considerable, with year-round growth, but with such cover the vegetation at ground
level is reduced. The herbivores that inhabit the tropical lowland rain forest are
browsers, while the predators are solitary hunters. Tall trees naturally provide attractive opportunities for climbing species to seek refuge as well as food, and it is no acci-
dent that many of Africa's monkeys, as well as its apes, are found in or near the
forested areas.

The eastern montane rain forests show a similar broad pattern of tall trees,
although the growth is less spread across the landscape and more limited by altitudi-
nal factors and the aspect of the slopes. The savanna gives way to rain forest at about
1500 m, but the forest cannot cope with conditions above around 2500 m and so typically yields to mountain bamboo belts. These, in turn, give way at around 3000 m to
subalpine moorland or heathland and at about 4000 m to alpine vegetation of quite
bizarre giant lobelias and senecio in short grassland.

FIGURE 2.11
Two aspects of tropical rain forest

(*Top*) A clearing in primary highland rain forest, Equatorial Guinea. (*Bottom*) A grove of palm trees of the genus *Raphia* in a lowland flooded area, Equatorial Guinea. (Photographs courtesy of Laureano Merino)

Wetlands

The swamplands, or wetlands, of Africa are mainly found inland and especially in the central regions, where drainage of the considerable amount of rain that may fall in the tropical highlands becomes choked and unable to reach the coast. Such areas are often where large bodies of water are believed to have existed. Parts of Sudan, Chad, Uganda, northern Zaire, northern Zambia, and Tanzania may have substantial wetlands, but perhaps the best known to visitors is the Okavango Delta of northwestern Botswana, where the Okavango River and its tributaries, arising in the highlands of western Angola and northern Namibia, run into the sands of the Kalahari Desert and produce a wetland area of almost 20,000 square km (figure 2.12). This provides a haven for such large herbivores as hippos, elephants, and buffalo, and of course attracts other species that normally inhabit the Kalahari during the drier periods that make desert life so difficult. In addition to swamplands proper are seasonally wet areas, or *dambos*, shallow-drainage grasslands often found at the upper end of plateau drainage systems, where seasonal waterlogging prevents the growth of trees. Good examples of *dambos* may be seen in the higher land surrounding the Bangweulu Swamps in northern Zambia. Such conditions are found in much of Africa coincident with the horseshoe-shaped distribution of broad-leaved woodland outside the central lowland rain forest, especially in the southern portion. O'Brien and Peters have suggested that such areas provide a reliable, year-round source of food for ungulates and predators, and would have been particularly attractive to early human groups.

Reconstructions of past vegetation and biomes are based on a number of lines of direct and indirect evidence. Identifiable plant remains sometimes are found, while fossil pollen may be recovered from suitable sediments. Photosynthesis by plants involves two major metabolic pathways, termed C3 and C4, that fractionate or alter the naturally occurring ratio of the stable isotopes ^{13}C and ^{12}C. Soils in areas with differing proportions of C3 and C4 plants—tropical and subtropical regions to about 40 degrees north and south of the equator, since C4 plants, mainly grasses, are restricted to those areas—may contain carbonates that preserve evidence of the vegetational differences. The teeth and bones of animals that ate the C3 and C4 plants record the differences, and the metabolic processes of the animals further fractionate the isotopic ratios. Thus it is possible to reconstruct the relative abundances of C3 and C4 plants, and hence the vegetation at a site, by analyzing the stable isotope compositions of soil carbonates and vertebrate remains and by combining that evidence with any traces of actual plants. A sufficient range of sites with such evidence or a

FIGURE 2.12
The Okavango Delta
A ground-level view of a wetland in the Okavango Delta of northwestern Botswana during the wet season. (Photograph by Mauricio Antón)

pollen sample, which probably derived from a regional pollen rain, can provide a picture of the flora and vegetation on a regional scale. More broadly still, the actual composition of animal assemblages found at a site can give a general idea of the habitat in the vicinity, although care must be exercised to avoid circularity by inferring vegetation from animal remains and then using the interpreted vegetation to support the picture derived from the animal remains.

Mammalian Distributions

Some general characteristics of African mammals may be seen across the continent as a whole. One gross feature of biogeography is the broad similarity at the family level of the mammalian faunas of Africa and the Oriental region—southern Asia in conventional zoobiogeographic terminology. This similarity reflects the recent presence of a tropical environment linking the two areas across the Arabian Peninsula before the opening of the Red Sea and the increasing aridification of the Middle East since the mid Pliocene. Indeed, plant biogeographers would conventionally combine the Oriental and Ethiopian regions into a single Paleotropical region.

The distribution of mammals in Africa reflects in many ways climatic and vegetational zones and hence geology and soil. The savannas, deserts, and forests have their distinct faunas, but some species occur in more than one biome, and the extent of habitat mosaicism allows a rich diversity of animals to coexist effectively within a region. The zoologist Peter Grubb has described the pattern of distribution as being characterized by regional faunas that are relatively discrete but show considerable overlap. He has divided High Africa into several biotic zones—the major ones being Northern and Southern Savanna, Somali Arid, Greater South-West Arid, and Central and Eastern Forest—and suggests a reasonable amount of overlap and exchange between drier and wetter savannas but a greater degree of isolation between arid zones when the total fauna is considered. However, large mammals in particular and eurytopic species in general have always been able to move freely between and across differing habitats, and large mammals are generally more diverse in areas closer to the equator, as has been usefully summarized by the zoologists Jane Turpie and Timothy Crowe.

The natural distributions of many of the large African mammals have been altered considerably by recent human activities, including farming, warring, hunting, and, of course, poaching. Nevertheless, it remains true that the savannas are home to one of the richest accumulations of mammals in the world, measured in terms of either overall biomass or diversity. Owen-Smith points out that large-herbivore biomass on fertile savannas may be up to three times that on infertile savannas and that the distribution patterns of those biomasses may vary greatly between the savanna types.

On the savannas, the seasonal growth of long, nutritious grasses and herbs in combination with the availability of woodland browse are two of the key factors in the establishment of biomass and diversity among large mammals. Taken overall, giraffes, zebras, hippos, rhinos, and warthogs occur together with a large proportion of the 72 species of indigenous antelopes, which include buffalo, eland, wildebeest, hartebeest, gazelles, impala, and dik-diks. The precise composition of the ungulate component varies considerably, however, especially in terms of the antelope species, and this variation is strongly correlated with the extent of woodland cover in the spectrum of savanna types from pure grassland to wooded grassland. The antelope tribes Antilopini and Alcelaphini, in particular, usually are associated with more open grassland and with grazing areas, in which they are joined by zebra and white rhino.

It is on the savannas, too, that the large carnivores operate to most effect, often in their biggest groupings. Lions and spotted hyenas are especially well adapted physically and behaviorally to group hunting and scavenging, as well as to protecting any food obtained. African hunting dogs, while active in such areas, are constrained by the appearance of hyenas, in particular, and may favor more closed terrain where hyenas are less able to hunt in large groups. Both spotted hyenas and hunting dogs have far more stamina than the cats, and a hunting group of either may pursue its prey for a considerable distance before bringing it down. Leopards and cheetahs maneuver around lions and hyenas by exploiting the favorable characteristics of the landscapes. Leopards employ cover and cache food in trees, while cheetahs undertake short, fast hunts of small to moderate-size ungulates that can be quickly consumed by an individual or a small family group before attracting the attention of larger competitors. That attention may be provoked by a competitor actually watching the hunt from close by, but seeing vultures circling overhead may also prompt it. These birds are constantly on the alert for carcasses, and, of course, their massed numbers in the sky are visible for a considerable distance.

A dominant feature of life and distribution patterns on the savannas is the need to cope with seasonal changes in rainfall and, therefore vegetation. During the wet season, the grass is lush and widely distributed away from watercourses; as the season progresses, good grazing becomes limited to damp areas, and herds must migrate. Evident on the savannas is grazing succession, whereby first zebras, for example, crop the coarser upper parts of the plants, to be followed by wildebeest, and finally by smaller animals like gazelles, which with their small mouths take the highly selective and more attractive lower parts. On the Serengeti Plain of Tanzania, an enormous migration begins once the dry season is under way in June, with some herbivores moving toward the woodlands in the west and north. Others may head toward the south, even entering the vast bowl of the Ngorongoro Crater and joining the resident population. The crater, the remains of a 2-million-year-old extinct volcano some 20 km across and with a floor area of around 250 square km, is home to the densest accumulation of prey species on the continent. The resident population is attracted by the year-round availability of water and therefore of fodder, mainly in the form of lush grasses. Buffalo are particularly abundant, although their numbers fluctuate over decades, and such a plethora of food produces an equivalent density of large predators. Lions and spotted hyenas coexist in a complex cycle of dominance, with first one and then the other operating as the major hunter from which the other tends to scavenge.

Despite the lower density of herbivores on the Serengeti in comparison with the floor of Ngorongoro, the actual number of animals that inhabit the plain are enormous, with perhaps 1 million zebras and wildebeest, as well as various gazelle species. With the onset of the rains in November, these animals move back to the grassland; during the migration, they must cross rivers swollen with floodwater and risk being swept downstream and taken by crocodiles.

Biomes other than savanna have their own special faunas. Life in the desert is highly conditioned by the need to cope with the extremes of temperature and the lack of water, and many animals display physical or behavioral adaptations related to these conditions. This is seen most notably in the addax antelope (*Addax nasomaculatus*), which can derive all the water it requires from vegetation and any dew on it. As a result of this ability, it was formerly found throughout the whole of the northern African gravel and dune deserts. In the southern African desert and in arid areas like the Namib, the Kalahari, and the Karroo, as well as in the dry lands of Sudan and Ethiopia, the oryxes or gemsboks, members of the genus *Oryx*, can also make do without free water if necessary. Other animals, such as the small desert fox, or fennec (*Fennecus zerda*), have huge, bat-like ears that no doubt serve both to help locate prey and to lose heat. Yet others, like the jerboas, members of the genus *Jaculus*, a small, long legged group of hopping rodents, build sophisticated burrows that maintain an internal microclimate.

Within the rain forests on either side of the equator, life is very different. High productivity in the vegetation is not necessarily matched by a great availability of foods for mammals on the ground, since much of the vegetation is at higher levels. Taï National Park in the Ivory Coast has 140 recorded mammal species, but more than 80 of them are rodents and bats. A relatively diverse group of browsers do find food, however, and animals like the okapi, as well as such antelopes as the bongo, buffalo, bushbuck, and numerous duiker species, are present, along with elephants, bush pigs and forest hogs, and the tiny chevrotain, a small ruminant herbivore superficially similar to a small antelope. The most common prey of local hunters is the small duiker, and outsiders may vastly underestimate the overall quantity of animal protein available in the rain forests. Most of the range of African monkeys are found here, too, especially the smaller monkeys that are able to reach fruits on the thinner branches where other, larger species cannot venture. Some, such as the colobus monkeys, which are able to leap across spaces between trees, may spend their entire lives above the ground level. Only the large adult gorillas are prevented by size from exploiting trees to any great extent, but juveniles and smaller females may still do so. But neither the diversity nor the biomass of large mammals of the rain forest rivals that of the more open areas. Cutting across all these biomes and biogeographic patterns is the distribution of one of the most cosmopolitan of African animals: the aardvark. This animal can be equally at home in habitats that range from dry savanna to rain forest, provided that sufficient termites are available and that burrows can be dug. In its eurytopic distribution, the aardvark resembles members of the order Carnivora, animals similarly constrained largely by the availability of food rather than precise conditions of climate or vegetation.

The island of Madagascar offers a microcosm of the range of African environments, with a humid eastern coast that supports rain forest, a cold elevated central region, a western belt of deciduous forest, and an arid southwestern area. Yet it supports only a fraction of the biota and the biotic diversity of the mainland, although it does have a large number of endemic species. Madagascar has long been separated from Africa and is home to only four mammalian orders— primates, rodents, insectivores, and carnivores—although pygmy hippos and a pig species became extinct during the Pleistocene. Most of its mammalian families are indigenous to Madagascar and rather primitive, but the antiquity of its separation from Africa implies that they must have been able to reach the island at some time during the early Cenozoic. How they did so is unclear because the Mozambique Channel, between Africa and Madagascar, is deep and no more modern mammals have managed to make the crossing. The most plausible explanation may be that the Comoros, a small group of islands between northwestern Madagascar and northeastern Mozambique, were once more extensive. In addition, recent geological evidence suggests that the Mozambique Channel at its narrowest, when the continent and the island were closest to each other, was marked by areas of dry land until 26 Ma ago as a result of compressive forces between the plates as the Indian subcontinent contacted the Eurasian landmass.

3 AFRICAN MAMMALS, PAST AND PRESENT

IN THIS CHAPTER, we present an outline history of each of the orders of large terrestrial mammals known from Africa and the range of species assigned to the various families. For completeness, we mention smaller mammals as well, although they usually have poor fossil records. For each family, we summarize the most important features of the living and fossil species, note their general distributions and habitat, and give an indication of size, with either a range of body weights or a single figure that is usually the upper weight limit for males.[*]

Methods of Reconstruction

One of the major aims of this book is to provide an accurate visual impression of the species that make up the changing large-mammal fauna of Africa and to set these species in an appropriate landscape. Modern animals present few problems for the illustrator, since live subjects, still photographs, and video footage of animals in motion are abundantly available, at least for some of the more common or generally better known species. Extinct animals present difficulties, however. Bones may

---

[*]We cannot, of course, duplicate the detail provided in a field guide to the modern species of Africa, and we recommend that readers in search of further information consult Jonathan Kingdon, *The Kingdon Field Guide to African Mammals* (San Diego, Calif.: Academic Press, 1997). A huge amount of behavioral information is also summarized in Richard Estes, *The Behavior Guide to African Mammals* (Berkeley: University of California Press, 1992), an equally excellent book.

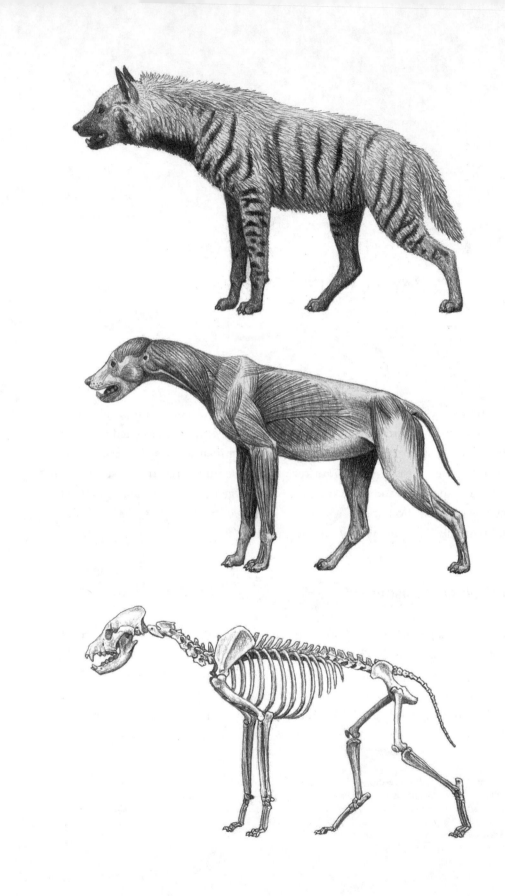

survive, but soft tissues such as muscle and skin, or impressions of them, do so in only the most exceptional conditions, while details like the color and patterning of the coat simply disappear. To add to the problems, individual skeletons are rarely complete, and certain parts of the skeletons of some species are still unknown. There is also an unfortunate tendency to give new species names to newly discovered or described material, often on the basis of fragmentary remains, leading in some cases to the absurd implication that certain taxa appeared at only one locality. The extent to which such newly proposed taxa differed morphologically, if at all, from more established ones is often far from clear, presenting a further problem for the illustrator.

The overriding rule that we have followed in reconstruction is that the life appearance of the animal must be firmly based, whenever possible, on the skeleton of the species, which provides a clear guide for the placement and bulk of the muscles. It is only after the muscles have been put in place that the external features can be added. Except for species whose skeletons are unknown, or virtually identical to those of living relatives, all the reconstructions in this book adhere to that rule, illustrated in figures 3.1 and 3.2. We start by drawing the assembled fossil skeleton in a life pose (figure 3.2) and then add the deep muscles, using comparisons with related living taxa as a guide (figure 3.1) but ensuring that the image of the animal built up at this stage is firmly based on skeletal information. The head, which embodies much of the "personality" of the animal, is similarly based on the shape and proportions of the skull, onto which we place the masticatory, or chewing, muscles. These define the main volumes of the head, although other features—such as the length of the nasal cartilage, the position and shape of the external ears and lips, and the presence of whiskers— also must be considered. The major unknown as far as life appearance is concerned then remains the color and patterning of the coat, and here we have generally been guided by the distribution of such features in living relatives. A more detailed reconstruction of our fossil hyena may be seen in plate 12.

This procedure enables us to provide a static reconstruction of the animals posed against a suitable background. But, of course, many of the characteristics of an animal are seen in the way it moves, whether it is a pig or an antelope running away from predators, a hyena or a cheetah chasing prey, or a giraffe in its characteristic splay legged stance drinking from a pool. When it comes to reconstructing movement, we can take the size and bodily proportions of the animal in question and apply many

FIGURE 3.1

Sequential dissection of *Hyaena hyaena*

The living striped hyena (*Hyaena hyaena*) serves as a good general model for the principles of reconstruction. Once the skin is stripped away, it is clear how the positioning and bulk of the major muscles relate to the underlying framework of the skeleton and thus how the appearance of the animal derives from that structure.

general principles from the study of living relatives or from simply similarly-shaped animals if the fossil form has no known descendants.

Terminology

One of the most important aspects of the terminology of paleontology is the system of naming animals and plants. A formal system of naming all the organisms on Earth, applicable to both living and extinct forms, was established more than 200 years ago by the Swedish naturalist Carolus Linnaeus (Carl von Linné) and is now universally used. A particular kind of animal, such as a chimpanzee or a brown bear, is known as a species, and a genus is a group of species thought to be more closely related to one another than to any other species. A species has a formal, two-part name composed of the generic and specific names. According to the Linnaean system, the generic and specific names are in Latin; the generic name begins with a capital letter and, after its first use, can be abbreviated to that letter; the specific name begins with a lower-case letter; and the whole species name usually is in italics. There are established rules of nomenclature, regulated by the International Commission on Zoological Nomenclature, and the issue of what is the correct name for a species can become immensely complex.

In this system, the lion is given the species name *Panthera leo*, consisting of the specific name *leo* and the generic name *Panthera*, and it is distinguishable by that name from the leopard (*Panthera pardus*) and the tiger (*Panthera tigris*). These three members of the genus *Panthera* are considered to be more closely related to one another than to the cheetah, which is placed in a separate genus and given the species name *Acinonyx jubatus*. To the nonspecialist, it may seem to be a clumsy way of naming an animal, but the species name is precise and, being in Latin, is comprehensible to all who know the system, no matter what their native languages (which, of course, have different common names for animals). It is particularly useful when taxonomists

FIGURE 3.2

Sequential reconstruction of *Ictitherium ebu*

The extinct hyena *Ictitherium ebu*, from the Miocene locality of Lothagam in Kenya, is reconstructed by reversing the procedure seen in figure 3.1. Although both animals have the same bones in the same place, the skeletal proportions of *Ictitherium ebu* differ from those of *Hyaena hyaena*: the fossil species has more gracile limbs, relatively longer hind limbs and therefore a more horizontal spine, a longer neck, and a differently shaped and less robust skull. Thus the muscle masses also differ, and the two species look different from each other in the flesh.

Reconstructed shoulder height: 60 cm. The estimates of shoulder height of extinct mammals given in this book are based on measuring a vertical line going from the ground to the top of the shoulder of an animal as reconstructed in a life-like, standing or walking pose. In contrast, the shoulder height of living mammals often is measured with a tape on the extended forelimb of a dead or an anaesthetized animal, sometimes providing slightly higher measurements.

deal with extinct animals that may have no common name in any language. Since there are about 2 million known and named living species of plants and animals, with a possible biota that may extend to 10 times that figure and an ever-expanding number of fossil taxa, the reasons for having a precisely agreed name for each is self-evident.

Various genera are, in turn, placed in families, such as the Felidae for the cats, the Canidae for the dogs, and the Hyaenidae for the hyenas. All three, together with other families like the Ursidae (bears), are, in turn, placed in the order Carnivora. Together with other orders—such as the Rodentia (rodents), the Artiodactyla (even-toed ungulates with split hooves, such as antelopes and deer), and the Primates (to which we belong)—the Carnivora make up the class Mammalia, which today consists of around 20 orders. This ranked system of naming, applied to the plant and animal kingdoms (a higher category still), reflects opinions about the closeness of the relationships between the various species, genera, and so forth, and therefore allows taxonomists to group them. As we shall see throughout the book, the higher categories in particular provide a useful shorthand terminology.

Thus we have a basic hierarchy of various taxonomic levels:

Kingdom	Animalia
Phylum	Chordata (Vertebrates)
Class	Mammalia
Order	Carnivora
Family	Felidae
Genus	*Panthera*
Species	*Panthera leo*

Our own species is classified in precisely the same way. Humans have the specific name *sapiens* and the generic name *Homo*, giving the species name *Homo sapiens*. We belong in the family Hominidae and the order Primates, so the bottom four places of our hierarchy is:

Order	Primates
Family	Hominidae
Genus	*Homo*
Species	*Homo sapiens*

There is a complication of terminology to be faced in the case of the Primates, however, and we shall examine the matter in a little more detail because doing so illustrates some of the general principles involved in taxonomy.

Taxonomy of the Primates

Traditional classifications would consider us and our closest fossil ancestors to be in the family Hominidae and to be separate from the great apes—the orangutan, chimpanzees, and gorillas—which would be in a separate family: the Pongidae. Both Hominidae and Pongidae would, in turn, be distinct from the family Cercopithecidae: the Old World monkeys, such as colobines and baboons. Clearly, the great apes are our closest living relatives, yet including the three great apes in one family is considered by many taxonomists to be illogical because the orangutan is thought to have diverged from the common ancestor of humans, gorillas, and chimpanzees before they diverged from one another. In other words, humans, chimpanzees, and gorillas share a *later* common ancestor not shared with the orangutan, so the orangutan, gorillas, and chimpanzees do not form a natural group that excludes humans. It would therefore make more sense to reorganize our positioning within the basic hierarchy. This places humans, near humans, and living and fossil great apes in the family Hominidae. It puts the orangutan in the subfamily Ponginae and humans, together with gorillas and chimpanzees, in the subfamily Homininae, with a further division between humans and the African great apes at what is known as the tribal level. Hominidae and other apes, such as the living gibbons and various primitive fossil apes, are then distinguished from the Old World monkeys at the superfamily level as Hominoidea. When we add other primates, including various fossil taxa, to this arrangement, we may then depict the order as follows:

Order   Primates
   Suborder   Strepsirhini (lemurs, lorises, and galagos)
     Superfamily   Lemuroidea (lemurs)
     Superfamily   Lorisoidea (lorises and galagos, or bushbabies)
   Suborder   Haplorhini (tarsiers, New and Old World monkeys, apes, and humans)
     Superfamily   Tarsioidea (tarsiers)
       Family   Tarsiidae (tarsiers)
       Family   Omomyidae (extinct tarsier-like primates)
     Superfamily   Ceboidea (New World monkeys)
     Superfamily   Cercopithecoidea (Old World monkeys)
       Family   Cercopithecidae (leaf monkeys, vervets, baboons, and macaques)
         Subfamily   Colobinae (leaf monkeys)
         Subfamily   Cercopithecinae (vervets, baboons, and macaques)
       Family   Victoriapithecidae (extinct early monkeys)

Superfamily   Hominoidea (apes and humans)
    Family   Proconsulidae (extinct primitive apes)
        Genus   *Proconsul*
    Family   Hylobatidae (gibbons)
    Family   Hominidae (great apes and humans)
      Subfamily   Dryopithecinae (extinct early hominids)
        Tribe   Afropithecini
        Tribe   Kenyapithecini
        Tribe   Dryopithecini
      Subfamily   Ponginae (orangutan and extinct relatives)
        Genus   *Pongo* (orangutan)
        Genus   *Sivapithecus*
      Subfamily   Homininae (African great apes and humans)
        Tribe   Gorillini (African great apes)
        Genus   *Gorilla* (gorillas)
        Genus   *Pan* (chimpanzees)
        Tribe   Hominini (humans and closest relatives)
        Genus   *Homo* (extinct and living humans)
          Species   *Homo sapiens* (living humans)
        Genus   *Australopithecus* (extinct early hominins)
        Genus   *Paranthropus* (extinct early hominins)

As may be seen, the inclusion of various other extinct and living and apes in the evolutionary picture is then achieved by giving them rank as different families within the Hominoidea (Proconsulidae and Hylobatidae) or subfamilies within the Hominidae (Dryopithecinae). Our own closest fossil relatives appear with us as species in the tribe Hominini, which may be rendered informally as hominin. This scheme also places the likely ancestors of the orangutan, members of the genus *Sivapithecus*, in the subfamily Ponginae. We shall explain the significance of the other taxa in this list—such as the family Proconsulidae, the subfamily Dryopithecinae, and the genera *Australopithecus* and *Paranthropus*—as well as the other taxa among the Old World monkeys, when we deal with the primates later in this chapter.

We have gone into some detail about the classifications in the order Primates because it is particularly important for clarity in discussions of human evolution and relationships within the order and, as we said, because it serves as a useful example of what lies behind the names and categories of all species. As may be expected, other orders and families of mammals have their own variously tangled and disputed systems of names and relationships. Although we shall mostly avoid getting side-tracked by those disputes, it is important to point out that decisions about species names and relationships, even of living organisms, are often far from straightforward. One authority may recognize five species of bushbaby; another, six. One may place mongooses (family Herpestidae) with civets (family Viverridae). For that reason, many writers often cite a particular authority as the source of their scheme of naming, although that is not always easy to do when dealing with living and fossil species across a very wide range of the biota. In order to avoid an overly complex text, we have avoided citing authorities. Although we have generally followed Jonathan Kingdon for many of the families of living African mammals, the names of fossil taxa are drawn from a very much wider set of sources.

Decisions about relationships between groups of living and fossil organisms, and even about the members of the groups themselves, have traditionally been based on morphology, the anatomy of an animal or a plant. But in the past few decades, the ability to look beyond such relatively gross patterns of morphology has developed with the growth of biomolecular analyses. These methods, it is argued, provide deeper insights since they can establish relationships on the basis of fundamental patterns in the arrangement of DNA, the building block of life. Such patterns may no longer be visible in the anatomy of organisms because they have been blurred by the animals' and plants' evolutionary histories since divergence. As with all new approaches, some—perhaps even many—of the results obtained are contentious and seem to conflict with what many morphologists feel are well-established patterns, and the arguments and detail lie mostly beyond the scope of this volume. However, the idea of an Afrotheria, a unique and previously unsuspected grouping of certain African mammals based on analyses of DNA sequences, does appear to have a strong foundation based on work by Ole Madsen and William Murphy and their respective colleagues. The Afrotheria group consists of elephants, hyraxes, tenrecs, aardvarks, elephant shrews, golden moles, and the aquatic manatees, or sea cows. The fact that the closest, or sister, group to these animals is the order Xenarthra—which consists of South American sloths, anteaters, and armadillos—makes sense in terms of continental drift and offers logical support to the identification of an Afrotheria.[*]

Order Primates: Lemurs, Lorises, Tarsiers, Monkeys, Apes, and Humans

The Primates, a diverse order with around 160 living species, probably originated before 65 Ma. They most likely evolved in the Northern Hemisphere, perhaps in

---

[*]Readers who wish to pursue these matters should consult the bibliography.

North America, where the so-called proto-primate genus *Purgatorius* is known from earliest Paleocene deposits, although Asia may be a stronger candidate for origins. If primates did originate so far back, then early forms were contemporaries of the latest dinosaurs. Africa has the greatest diversity of living primates, with around 25 genera and almost 70 species.

The order has been divided in various ways, and living species often have been placed in two suborders: the "lower" primates, the Prosimii (prosimians), and the "higher" primates, the Anthropoidea (simians, or anthropoids). The Prosimii includes the lemurs, lorises, galagos, and tarsiers—most of which are small and have a poor fossil record—and this suborder may have entered Africa as early as 55 Ma. The Anthropoidea consists of the New World monkeys of Mesoamerica and South America and the Old World monkeys and apes as well as humans. More recent views place the tarsiers closer to the Anthropoidea, however, and one suggested solution to the taxonomic problem that results from reclassifying the tarsiers is to redesignate the Anthropoidea as a hyperorder; create a second hyperorder, Tarsiiformes; and place both within the suborder Haplorhini. Lemurs, lorises, and galagos are then placed in the suborder Strepsirhini. It is the Anthropoidea that we shall concentrate on here after a preliminary review of the Strepsirhini and the Tarsiiformes.

FIGURE 3.3
Ring-tailed lemur

The ring-tailed lemur (*Lemur catta*) is one of the most popular species of the family Lemuridae. It is a diurnal, social, and largely terrestrial primate that, like the other members of its family, dwells in the forested areas of Madagascar.

Superfamily Lemuroidea: Lemurs

The lemurs are a superfamily of around 20 species of small to moderate size placed in four families—Lemuridae (true lemurs), Indridae (larger-bodied lemurs), Cheirogaleidae (dwarf lemurs), and Daubentonidae (aye-aye)—confined to the island of Madagascar. A typical specimen, the ring-tailed lemur (*Lemur catta*), is shown in figure 3.3. With the exception of *L. catta*, which can, of course, move perfectly well in trees, all lemurs are essentially arboreal and subsist on fruit and leaves. Fossil lemur-like primates, placed in the family Adapidae and genus *Adapis*, are known from the Eocene of North America and from deposits of 35 Ma at Quercy in France, but the history of the group in Africa is currently unknown. Among the subfossil species known from Madagascar are some giant forms.

Superfamily Lorisoidea: Lorises and Galagos

The lorises or slow lorises, also generally known as pottos (*Perodicticus potto*) or angwantibos (*Arctocebus calabrensis* and *A. aureus*), are small, slow-moving arboreal inhabitants of the western central African forest. As did those of the lemurs, fossil ancestors of the lorises seem most likely to have come from among the Adapidae.

The galagos, or bushbabies (figure 3.4), are an African group with several species in perhaps five genera: *Otolemur*, *Euoticus*, *Galago*, *Sciurocheirus*, and *Galagoides*. Tiny to small (up to 2 kg), they are long-tailed nocturnal creatures with elongated hind legs and enormous eyes, found in wooded areas of central and southern Africa.

Superfamily Tarsioidea: Tarsiers

The tarsiers, belonging to the single genus *Tarsius*, are large-eyed nocturnal primates now confined to woodlands of southeastern Asia (figure 3.5). They are small animals, weighing less than 200 g, with elongated tarsal bones, from which their name derives (Latin, *tarsus* [ankle]), used for leaping from tree to tree. Although lacking some anthropoid characteristics, such as an enlarged brain relative to body size, they have other features, such as a dry snout and reduced nasal region, that link them morphologically to the higher primates and are shown by molecular biology to be closely related. Although tarsiers do not live in Africa, fossil tarsier-

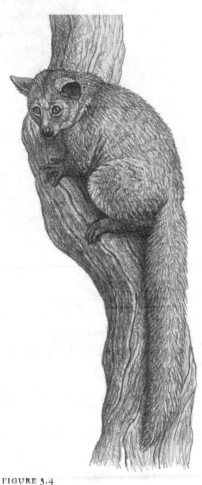

FIGURE 3.4
Thick-tailed bushbaby

The thick-tailed bushbaby (*Otolemur crassicaudatus*) is the largest of the galagos, and it moves along the branches of trees in a slower and more deliberate manner than its smaller cousins, although it can leap efficiently if necessary. It is a nocturnal primate that forages both on the ground and on the branches for its diet of fruit, gum, and small animals.

FIGURE 3.5
Western tarsier

The Western tarsier (*Tarsius bancanus*), like all the members of its genus, is a diminutive primate that lives in the forests of tropical Asia. Tarsiers get their name from their elongated tarsal, or foot, bones, which are part of their adaptation to move with impressive leaps among the branches of trees. Their enormous eyes are immobile, but the animals can rotate their heads nearly 180 degrees in each direction to compensate.

like animals placed in the genus *Afrotarsius* are known from the deposits of the Fayum Depression in Egypt, dated to 35 Ma. These animals may derive from an earlier family, the Omomyidae, known from deposits of Eocene age in Europe and North America.

Superfamily Cercopithecoidea: Old World Monkeys

Family Cercopithecidae: Leaf Monkeys, Vervets, Baboons, and Macaques
The Cercopithecidae, the largest primate family with over 70 living species, may be divided into subfamilies in order to accommodate both living and extinct forms.

SUBFAMILY COLOBINAE: LEAF MONKEYS
The Colobinae include the acrobatic arboreal species, such as the central and eastern African monkeys of the genera *Colobus* and *Pliocolobus*. They have complex stomachs and eat leaves—hence the name leaf monkeys, which often is given to them—and either lack thumbs or have almost lost them, a feature from which their name derives (Greek, *colobe* [cripple]). The resulting hand is well suited to use as a hook while moving through the trees but makes the manipulation of objects difficult and points to

FIGURE 3.6

Life reconstruction of *Paracolobus chemeroni*, compared with *Colobus guereza*

A remarkably complete articulated skeleton of *Paracolobus chemeroni* was found at the Pliocene site of the Chemeron Beds near Lake Baringo, Kenya, providing an unusually accurate picture of the monkey's body proportions. The fossil skeleton shows the typical features of colobine monkeys, with a dentition adapted to chewing leaves and limb bones indicative of arboreal locomotion. *Paracolobus* was much larger than any living African colobine, as is evident when it is shown to scale with the extant Guereza colobus (*Colobus guereza*).

a long history of arboreal specialization. Figure 3.6 shows examples of living and fossil colobines.

SUBFAMILY CERCOPITHECINAE: VERVETS, BABOONS, AND MACAQUES

The Cercopithecinae, or cheek-pouched monkeys, include the smaller monkeys like the vervets (figure 3.7) and the more terrestrial baboons, members of the genera *Papio* (figure 3.8) and *Theropithecus* (figures 3.9 and 3.10; plate 9). The fossil record shows considerable diversity, but is patchy; many cercopithecids simply have not lived in habitats where their remains were likely to have been incorporated into deposits. An early representative, *Propliopithecus*, a genus of uncertain affinities perhaps best placed in a separate extinct family, the Propliopithecidae, is known from the deposits of the Fayum Depression in Egypt, dated to around 33 Ma, but a large gap then appears in the record until around 17 Ma. At that point, several obvious cercopithecids are known, particularly from Kenya, and usually are referred to a separate fam-

FIGURE 3.7

Vervet monkey

Although it moves efficiently on the ground, the vervet monkey (*Cercopithecus aethiops*) is less well adapted to terrestrial locomotion than is the baboon, its relative in the subfamily Cerco-pithecinae. As demonstrated by a walking vervet in Samburu National Park, Kenya, vervets rest the entire palm of their forefeet on the ground, which is termed palmigrady, while baboons are more digitigrade, implying that they usually walk on their fingers only. Both monkey species are planti-grade on their hind feet, though, as seen clearly in this vervet. (Photograph by Mauricio Antón)

FIGURE 3.8

Olive baboon

All baboon species have fearsome canine teeth, especially in the males, and these teeth are used extensively in display and aggression. This male olive baboon (*Papio anubis*) in the Masai Mara National Reserve, Kenya, displayed its impressive fangs to warn the photographer not to get any closer, a warning that was immediately heeded! (Photograph by Mauricio Antón)

FIGURE 3.9

Sequential reconstruction of the head of *Theropithecus brumpti*

Well-preserved skulls and mandibles of the fossil baboon *Theropithecus brumpti* have been found in Pliocene deposits in the Shungura Formation at the Omo River, Ethiopia, and the Koobi Fora Formation at East Turkana, Kenya. Like the modern gelada baboon (*T. gelada*), the only living species of the genus, *T. brumpti* had robust cheek teeth adapted to chew tough vegetation and very strong masticatory muscles. But it had a longer muzzle, with impressive canines that no doubt were used for display by the males, which were very large in this sexually dimorphic species. In order to keep a powerful enough bite in spite of the long muzzle, the insertion areas of the masseter muscles in the skull of *T. brumpti* migrated forward, and a visor-like projection developed in the malar bones to anchor the most anterior fibers of the muscle. In this way, the masseter could transmit enormous pressure to the mandibular teeth, even to those positioned more anteriorly on the long jaw. It has been suggested that this adaptation enabled *T. brumpti* to break large-diameter objects, such as nuts, with the anterior premolars. This development is similar in some ways to that found in the "robust" australopithecines of the genus *Paranthropus*, which have been given the nickname nutcracker men.

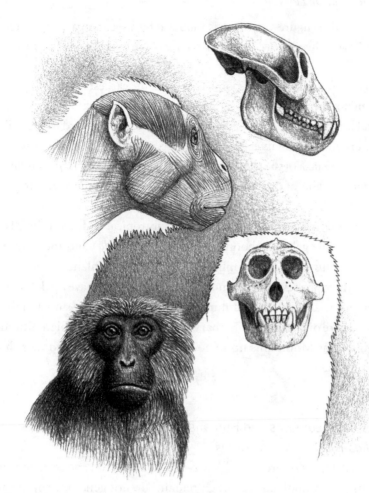

FIGURE 3.11

Sequential reconstruction of the head of *Victoriapithecus maccinnessi*

This reconstruction is based on an especially well preserved skull of the extinct monkey *Victoriapithecus maccinnesi*, discovered in Miocene deposits at Maboko Island in Lake Victoria, Kenya, and described in 1997. The skull shows that this monkey had unusually well developed nuchal and sagittal crests, in turn indicating the presence of very strong temporalis muscles. In other respects, the skull is typically cercopithecine-like, and the teeth suggest a mostly frugivorous diet. The animal would have weighed about 4.5 kg in life, and the total length of the skull is about 12 cm.

FIGURE 3.10

Sequential reconstruction of *Theropithecus oswaldi*

Skulls, partial skeletons, and isolated bones of the fossil baboon *Theropithecus oswaldi* have been found in Pleistocene deposits in the Koobi Fora and Kanjera Formations and at Olorgesailie, all in Kenya, among other sites. Like *T. brumpti*, this monkey was sexually dimorphic, and the males could attain enormous proportions, reaching the size of a female gorilla. The skeleton of *T. oswaldi*, like that of the modern gelada baboon (*T. gelada*), shows clear adaptations for terrestrial locomotion. It further resembles the gelada in details of the anatomy of the hand and the hind limb that indicate the habits of picking up small objects (seeds, short grass, or leaves) between index finger and thumb and of "shuffling" bipedally while foraging for such food items on the ground.

Reconstructed shoulder height: 78 cm.

ily, the Victoriapithecidae (figure 3.11). This is followed by a second gap until after 5 Ma, when a considerable diversification seems to have occurred involving numerous baboon species.

The paleontologist Brenda Benefit has pointed out that the earliest known monkeys appear to have been not the arboreal, leaf-eating animals that we might have expected, but semiterrestrial animals with dental adaptations to feeding on hard objects like seeds. A trend toward the loss of larger-bodied animals seems to have taken place, so only half a dozen baboon taxa of the genus *Papio* and a single species of *Theropithecus* exist today, with body weights of up to 50 kg. The rest of the African monkeys are generally small to medium size, with body weights usually well below 20 kg.

The nonhuman primates in general are tropical or subtropical in distribution. The African exceptions are the gelada baboon (*Theropithecus gelada*) and the Barbary macaque (*Macaca sylvanus*). The geladas inhabit the Ethiopian highlands, at altitudes of 2000 to 4000 m, where grasses make up much of their diet. They have thick fur and a complex pattern of social behavior that minimizes disruptive aggression and thus unnecessary expenditure of energy. The Barbary macaques occupy the high-altitude cedar forests of the Atlas Mountains up to 2000 m and have thickened hair that appears almost wool-like.

Superfamily Hominoidea: Apes and Humans
Family Proconsulidae: Extinct Primitive Apes
Early apes appear in the fossil record of Africa between 30 and 25 Ma and seem to have been adapted to tropical woodland. Although remains do not generally survive to become fossils in such a habitat, fossil apes have been found in Africa, and there is a reasonably good record from the late Oligocene to the mid Miocene, around 27 to 12 Ma.

The first to appear are placed in the Proconsulidae, a family that takes its name from the genus *Proconsul*, a range of primitive-looking African fossil apes (figure 3.12). Other genera allocated to this family and part of what we might term the proconsulid radiation are *Rangwapithecus* and *Nyanzapithecus*, and all are classified as hominoids by virtue of such characters as a relatively large brain, tooth morphology, and the absence of a tail. Members of the family vary in size from that of a large monkey to that of a female gorilla. They are found at a number of early Miocene sites in Uganda and in Kenya, such as Rusinga Island in Lake Victoria, Koru, and Songhor, where deposits are dated to 22 to 17 Ma. One of the oldest specimens comes from Lothidok Hill in northern Kenya, in deposits that may be as early as 27 Ma. The proconsulids are generally best characterized as arboreal quadrupeds; that is, they share generalized morphological features of probable monkey-like ancestors, with a four-

FIGURE 3.12

Life reconstruction of *Proconsul africanus*

This reconstruction, based on cranial and postcranial remains from Rusinga Island in Lake Victoria, Kenya, shows *Proconsul africanus* as a quadrupedal, partly arboreal primate. The shape of the articulations in its limb bones suggests that its locomotion would have resembled that of a macaque or even a New World monkey more than that of an extant hominoid.

Reconstructed shoulder height: 45 cm.

footed pattern of mobility placing them at home in the trees despite their ape-like grade of evolution, as evidenced by other characters.

Family Hominidae: Great Apes and Humans

Humans have long been classified as primates, and the degree of morphological similarity between humans and the great apes clearly points to a close relationship. Indeed, as outlined earlier, humans are included in the Hominidae, the same family as the orangutan, gorillas, and chimpanzees, and thus are, in effect, simply apes. Further evidence for this closeness, and for the scheme of relationships outlined in the taxonomic chart, is provided by studies of the basic genetic information in our cells, DNA, that suggest that we share around 98 percent of DNA with the African apes. It is then possible to go one step further and to compare the DNA of different species and to estimate the time of divergence from a presumed common ancestor.

The methods of doing so are complex, and the results to date are probably not the final answer. But comparisons of DNA derive estimates of a separation of the orangutan at about 12 to 10 Ma and a split between hominins and gorillins—that is, between humans and African great apes—at about 7 to 5 Ma. This relatively independent estimate of times of separation may be compared with the fossil record. We say "relatively" because, of course, the estimate still has to be calibrated to an event of "known" date, usually something else from the fossil record, but we know that we should be looking for evidence of a common ancestor of humans and African apes at about that latter time.

SUBFAMILY DRYOPITHECINAE: EXTINCT EARLY HOMINIDS

The stage after the Proconsulidae is a further radiation between around 17 and 12 Ma expressed by the division of the Dryopithecinae into the tribes Afropithecini, Kenyapithecini, and Dryopithecini.

Tribe Afropithecini

The Afropithecini are known from only Africa and Arabia, at localities dated to around 15 Ma, such as Maboko Island in Lake Victoria and Buluk, both in Kenya, and Ad Dabtiyah in Saudi Arabia. Material from Maboko in particular, earlier tentatively referred to the genus *Kenyapithecus* as *Kenyapithecus africanus*, is now recognized on the basis of new material from mid Miocene deposits in the Tugen Hills of Kenya as belonging to a new genus, *Equatorius*.

The afropithecines are characterized by thick tooth enamel and premolars larger than those of the proconsulids, which indicate a possible change to harder food. The skeleton is little known, but again a generalized tree-dwelling adaptation using all four limbs is indicated by material from Maboko Island. Thus little change over the more primitive proconsulids is evident.

Tribe Kenyapithecini

The Kenyapithecini are known from Fort Ternan in Kenya, dated to around 14 Ma, but the majority of the later specimens come from sites in Turkey and southeastern Europe. The kenyapithecines seem little advanced over the afropithecines.

Tribe Dryopithecini

The Dryopithecini are European in distribution and appear to have inhabited subtropical to warm temperate forests, where the seasons were somewhat marked and at least part of the vegetation may have been deciduous (figure 3.13).

FIGURE 3.13
Life reconstruction of *Dryopithecus laietanus*

This reconstruction, based on remarkable fossils from Can Llobateres in Spain, shows *Dryopithecus laietanus* as a relatively primitive hominoid, but one that already shows incipient orangutan-like features, many of which are adaptations for an arboreal locomotion that included a high proportion of brachiation.

SUBFAMILY HOMININAE: AFRICAN GREAT APES AND HUMANS

Tribe Gorillini: African Great Apes

The common chimpanzee (*Pan troglodytes*), pygmy chimpanzee or bonobo (*P. panis-cus*), and gorilla (*Gorilla gorilla*), members of the tribe Gorillini, have central African distributions. Typical specimens of a common chimpanzee and a gorilla are shown in figure 3.14. The most complex distribution is that of the gorilla, with large western

FIGURE 3.14

Living African great apes

The stance of the common chimpanzee (*Pan troglodytes*) (*left*) underlines the basic quadrupedalism of all the apes, while the walking western lowland gorilla (*Gorilla gorilla gorilla*) (*right*) shows the hind limbs and bent-legged posture that renders such movement possible when necessary but highly inefficient. The depiction of the bipedal locomotion of the gorilla is based on footage of wild specimens from the Republic of Congo in central Africa.

and smaller eastern lowland populations separated by the central Congo Basin and a very small highland population in Rwanda and Zaire. Various taxonomic schemes define the populations as subspecies or, in the case of the western lowland group, a separate species. All are threatened, but the western population is currently rather large. Gorillas form fairly small social groups dominated by a mature male, an animal that may weigh well in excess of 200 kg and be twice the size of a female.

The common chimpanzee, a much smaller and less dimorphic species than the gorilla in which males weigh up to around 40 kg, is found in woodland areas from Senegal eastward to western Uganda and Tanzania and bounded to the south by the Congo River. Chimps have complex social lives and live in large communities of up to 100 animals, with males defending the territory aggressively. One marked feature of behavior is the hunting of monkeys, often undertaken by groups acting cooperatively. The Congo River is the northern limit of the bonobo's range. The bonobo is slightly smaller than the common chimpanzee, with longer legs and a rounder skull.

It is markedly less aggressive, with a more dispersed population structure. Humans in the Congo Basin consider the bonobo a good source of food

The evolutionary histories of the living African great apes are easily dealt with, since they have no known fossil record (although the *Ardipithecus ramidus* material, discussed later, has been said to show similarities to the chimpanzee). This rather astonishing fact may be in part accounted for by their preference for wooded environments. Although, as we have seen, the African record of apes does include animals with arboreal adaptations, it may be that the pattern evident in the fossil record is generally biased toward those taxa and lineages with rather more generalized habitat preferences than those of the gorillins.

What this means in practice is that we have no current knowledge of the immediate common ancestor of humans and the African great apes—it is not clear what relationship, if any, they have with the various tribes of the Dryopithecinae. The skeleton of the living African apes does have some similarities with that of *Dryopithecus fontani*, which could be taken to imply that the apparent African origins of humans and apes owes itself to a later Miocene dispersion from Eurasia, but the similarities may be misleading.

Tribe Hominini: Humans and Closest Relatives

Since the human lineage originates in Africa, it is evident that one or more dispersions of members of the tribe Hominini must have taken place from there. Our own genus, *Homo*, is conventionally considered to be evident in Africa back to about 2.5 Ma, first represented in eastern Africa by the species *Homo habilis* (figure 3.15) and *H. rudolfensis* (figure 3.16). Stone tools appear in the archaeological record at about the same time, leading to speculations about the relationship between evolutionary change and the development of toolmaking abilities. However, it has long been apparent that the earliest taxa assigned to the genus *Homo* are rather heterogeneous, even when split into several groups, and arguments that they be removed from *Homo* altogether have recently reemerged in the work of Bernard Wood and Mark Collard. If that is done, the earliest known representative of our genus would be the material variously referred to as *H. erectus* or *H. ergaster* (plate 10). This species is known in East Africa from at least about 1.8 Ma, based on the cranium KNM-ER 3733 from the Koobi Fora Formation, on the eastern shore of Lake Turkana in Kenya. The best preserved specimen of all, and one of the most complete early hominins by far, is the skeleton of a youth from the western side of Lake Turkana (KNM-WT 15000), found in deposits dated to 1.5 Ma (figure 3.17).

Homo erectus appears to have been a fairly long-lived species, existing for around 1 Ma. The next hominin species to be recognized in the African record is now generally termed *H. heidelbergensis* (the type specimen of the species coming from a site

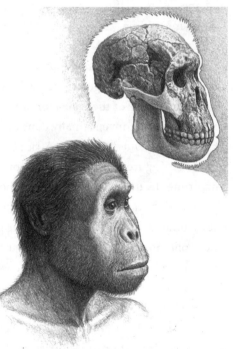

FIGURE 3.15

Life reconstruction of *Homo habilis*

The body proportions in this reconstruction of *Homo habilis* are based on a partial, probably female skeleton (OH 62) from Olduvai Gorge in Tanzania, while details of the morphology of the skull, hands, and feet are based on unassociated material from Olduvai and from East Turkana in Kenya. The Olduvai female was a diminutive hominin about 1.1 m tall, with strikingly long arms and short legs. These proportions and several features in the shape of the limb bones indicate a climbing ability superior to that of modern humans, but the shape of the feet and other traits indicate that *H. habilis* already was an obligate biped that would have spent most of its time on the ground.

FIGURE 3.16

Life reconstruction of the head of *Homo rudolfensis*

This reconstruction of the head of *Homo rudolfensis* is based on the cranium KNM-ER 1470, from East Turkana in Kenya. The cranium displays a mixture of traits that makes its classification difficult. The hominin has a primitive look, resembling *H. habilis*, but its brain is large, its face is long and flattened, and the crests for muscular insertions are rather faint. The view chosen for this reconstruction clearly shows how long and flat the face was, and how moderate the supraorbital ridges.

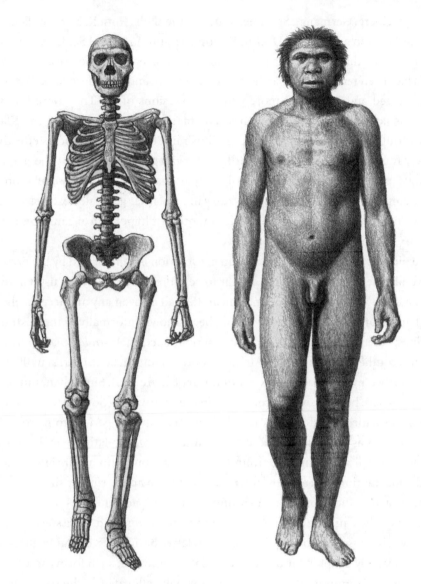

FIGURE 3.17

Life reconstruction of the "Turkana boy"

This reconstruction shows the "Turkana boy," represented by a semicomplete skeleton from Nariokotome at West Turkana, Kenya, as he might have looked at the time of his death, when he would have measured about 1.6 m. It has been estimated that if he had reached adulthood, he would have grown to a height of 1.8 m. The unusual preservation of this specimen allows the accurate reconstruction of body proportions, showing that he already had relatively long legs and short arms, much as in modern humans. More problematic is the reconstruction of the pelvis in frontal view. It was originally proposed that the pelvis was narrow, as in modern humans, rather than having the widely flaring illia of australopithecines, but the bone is damaged and the interpretation open to question. The later discovery of a complete pre-Neanderthal pelvis from Atapuerca in northern Spain showed that wide illia probably were the wide-spread condition among early hominins, and it seems likely that a narrow pelvis appeared only with *Homo sapiens*. In this reconstruction, the Turkana boy has a wide pelvis, based largely on a specimen from East Turkana (KNM-ER 3228) that may belong to an early *H. ergaster* or *H. erectus*.

near Heidelberg, Germany). Specimens such as the skulls from Kabwe (Broken Hill) in Zambia, Omo 2 and Bodo D'Ar in Ethiopia (figure 3.18), and Saldanha in South Africa make up this species, with a date of around 500,000 to 200,000 years ago. These fossils had been referred to as "archaic" *Homo sapiens*, and do indeed look like rather more rugged and somewhat more primitive versions of us. They were followed around 150,000 years ago by a group based on material from Border Cave and Klasies River Mouth in South Africa and Omo-Kibish in Ethiopia that appears effectively entirely modern. These fossils, the earliest known appearance of the modern morphology, together with results of genetic analyses that indicate the close relationship of all living humans and a common ancestry in Africa, has generated extensive media interest in recent human origins. Inevitably, our last hypothetical common female ancestor has been named the African Eve.

The picture of the earlier stages of human evolution is complicated by the presence of several other extinct forms that appear to be more closely related to us than to the African apes, although that does not mean that all or even any are necessarily our direct ancestors. Many of them have long been known informally as the australopithecines, based on the genus of the first described species, *Australopithecus africanus* (literally, southern ape-man from Africa), found at Taung in South Africa in the 1920s and described by the anatomist Raymond Dart (figure 3.19). Since then, numerous specimens have been recovered from caves in the Transvaal of South Africa and from open-air localities in eastern Africa, with dates spanning the period from around 4.5 Ma to perhaps as late as 1 Ma. The australopithecines were relatively small, no more than 150 cm tall and weighing around 30 to 50 kg, but with an upright posture, bipedal locomotion, and a brain larger than that of an ape of similar size.

Much taxonomic debate has surrounded the australopithecines, including the question of their precise relationship to the genus *Homo* and the issue of human ancestry. The close relationship is based on relative brain sizes, dental features, and skeletons largely adapted for upright, bipedal locomotion; such forms accordingly belong in the Hominini. If the specimens formally allocated to the species *Homo habilis* and *H. rudolfensis* are removed from the genus *Homo*, they might most logically be classified as australopithecine species.

All investigators have generally recognized two kinds of australopithecine: a more-lightly built one, often known as "gracile," and a more rugged form, commonly referred to as "robust." Some have placed the robust forms in a separate genus, *Paranthropus*, with up to four species identified: *Paranthropus robustus* (figure 3.20) and perhaps *P. crassidens* from southern Africa, and *P. boisei* (figure 3.21) and *P. aethiopicus* (figure 3.22) from eastern Africa. The robust taxa are known from around 2.2 Ma to perhaps as late as 1 Ma. They are distinguished from the gracile forms not so much by being slightly larger, standing between 110 and 150 cm and

FIGURE 3.18

Life reconstruction of the head of *Homo heidelbergensis*

The mandible of this very robust specimen was restored on the basis of fossils of *Homo heidelbergensis* from Europe and of *H. ergaster* from Africa. The well-preserved skull, from Bodo D'Ar in Ethiopia, is very large and robust, and in life the face would have reflected those proportions. The surface of the bone shows scratch marks, indicating that this individual was defleshed using stone tools, perhaps an act of cannibalism.

weighing between 30 and perhaps 50 kg, as by having extremely robust skulls and large back teeth. Interestingly, some of the largest fossil specimens known have actually been referred to the earlier species *Australopithecus afarensis*. The morphology of the skull and the dentition of the robust forms point to a considerable emphasis on chewing with powerful muscles. All suggestions about their relationship to *Homo* tend to regard the robust forms as a side shoot of the main human lineage.

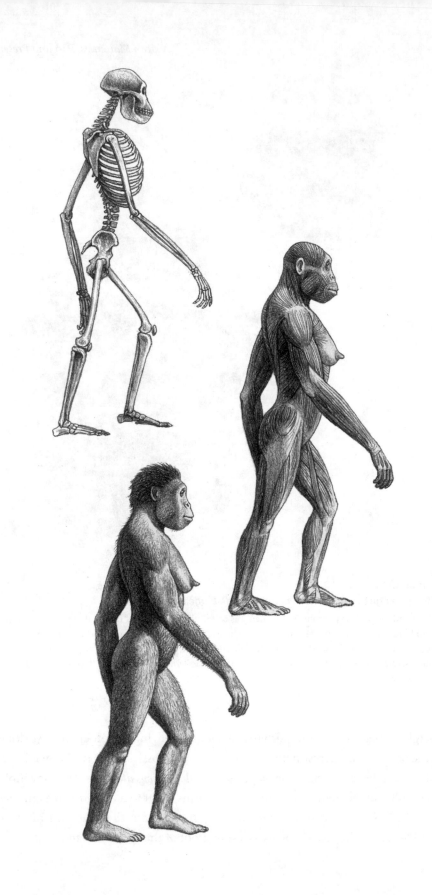

Life reconstruction of the head of *Paranthropus robustus*

The cranium and mandible of a female *Paranthropus robustus* from Drimolen Cave, in South Africa, nick-named Eurydice by its discoverers, are the most complete and well preserved of a paranthropine found to date. The shape of the cranium is remarkably similar to that of a partial skull from East Turkana in Kenya, attributed to a female *P. boisei*. It is considerably smaller than skulls attributed to male *P. robustus*, and the nuchal and saggital crests are much less developed, indicating the presence of marked sexual dimorphism in this species.

Sequential reconstruction of *Australopithecus africanus*

This reconstruction of *Australopithecus africanus* combines remains of several individuals from the site of Sterkfontein in South Africa, including the well-preserved skull STS 5 ("Ms. Ples") and the excellent specimen STS 14, which comprises a partial vertebral column, pelvis, and femur. These fossils give a clear idea of posture and general morphology, but leave some uncertainties about the relative proportions of the limbs, which would appear to be similar to those of *A. afarensis* but perhaps with relatively longer arms. The study of an almost complete skeleton currently being excavated at Sterkfontein probably will solve some of these uncertainties and add much more information to the record.

FIGURE 3.21

Life reconstruction of the head of *Paranthropus boisei*

Cranial remains of *Paranthropus boisei* from Olduvai Gorge in Tanzania and the Koobi Fora Formation at East Turkana, Kenya, provide a good picture of morphology and even variability in the skull, but hardly anything is known about the postcranial skeleton, which is assumed to have been generally similar to that of *Australopithecus afarensis*. The head of these hominins would have been rather impressive in life because of the great development of the masticatory apparatus, particularly the anterior projection of the insertion areas for the masseter muscles in the malar bone, which made the whole face look flat and even slightly concave (see figure 5.13).

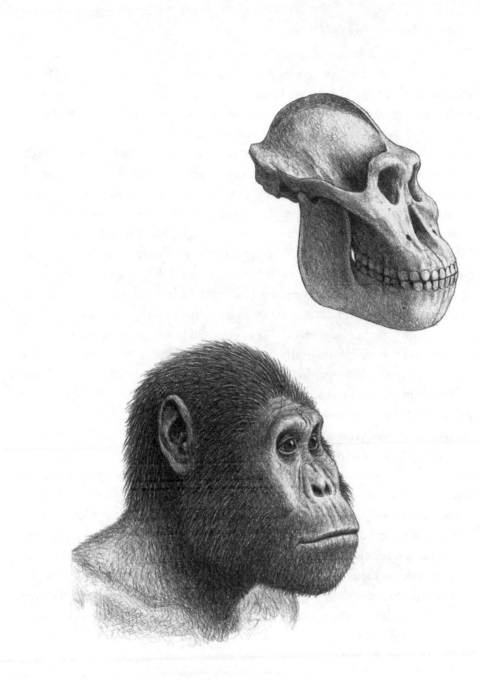

FIGURE 3.22

Life reconstruction of the head of *Paranthropus aethiopicus*

This reconstruction of the head of *Paranthropus aethiopicus* is based on the so-called Black Skull (KNM-WT 17000), from West Turkana in Kenya. This skull combines some primitive features of *Australopithecus afarensis* with derived traits typical of *P. boisei*, and thus seems a likely candidate to be an ancestor of that species and maybe of *P. robustus*. Its enormous saggital crest indicates the presence of exceptionally well developed temporalis muscles.

The number of gracile australopithecine species is the subject of much debate. They include the South African type species, *Australopithecus africanus*, and the East African *A. afarensis* (figure 3.23). The latter is based on material found in the 1970s at localities at Hadar in Ethiopia and Laetoli in Tanzania, with an age between 3 and almost 4 Ma, and until recently was the oldest known hominin. The *A. afarensis* sample includes the widely known, semicomplete skeleton of Lucy, a female around 106 cm in height and 30 kg in weight. It may also contain the material from deposits of similar age found at Koro Toro in Chad in 1993, although the discoverers have suggested that this form may belong to a separate species: *A. bahrelghazali*. Males of the species were considerably larger than females, perhaps 45 kg in weight. Specimen AL 442-2, from Hadar, is the largest australopithecine cranium known, despite a cranial cavity (and therefore crude estimate of brain size) of only 500 cc. Studies suggest a brain larger than that of an ape of comparable size and an upright, bipedal way of walking, but the method of locomotion differed in details from that of modern humans and was combined with some considerable ability to climb and move about in trees. As both the oldest and the morphologically most primitive of the australopithecines, *A. afarensis* has been widely regarded as a good general ancestor of all later hominins, including humans. Recent discoveries in Chad appear to have extended the known distribution of the species westward by around 2000 km into central northern Africa.

Six recent discoveries have further complicated the picture, however. The first consists of fragmentary material from East Turkana and Kanapoi in Kenya, found in deposits with dates around 4.2 to 3.9 Ma, that has been placed in a new species: *Australopithecus anamensis*. This species, for which detailed analyses of its functional morphology are not yet available, is considered to be even more primitive than *A. afarensis*. The second discovery consists of equally fragmentary and yet more primitive specimens from Aramis in the Middle Awash Valley of Ethiopia, dated to around 4.4 Ma. First descriptions of the fossils referred them to a new species, *A. ramidus*, but the authors of that name later suggested that the species be placed in a separate genus as *Ardipithecus ramidus*. They also pointed to similarities to chimpanzee morphology in various features of the remains. Further fragmentary material from the Middle

---

FIGURE 3.23

Life reconstruction of *Australopithecus afarensis*

This reconstruction incorporates information from fossils of several individuals of *Australopithecus afarensis* from the site of Hadar in Ethiopia. The relative body proportions follow those of the famous female skeleton nicknamed Lucy, while absolute size and skull morphology are based on fossils of larger specimens considered to be males. In spite of its greater geological age and generally more primitive features, *A. afarensis* appears to have had somewhat more "modern" body proportions, with longer legs and shorter arms, than *Homo habilis*.
Reconstructed total height: 1.45 m.

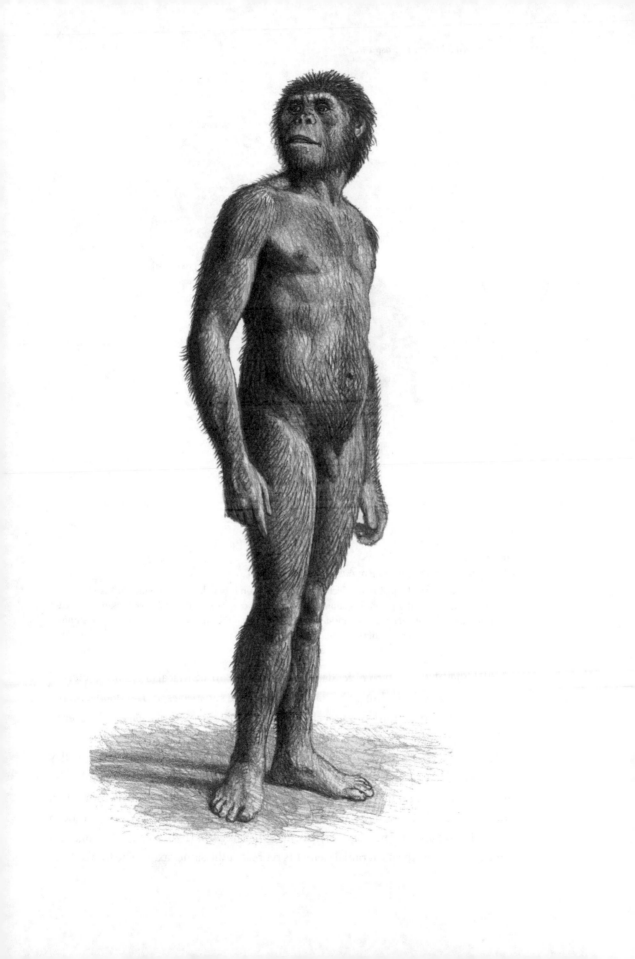

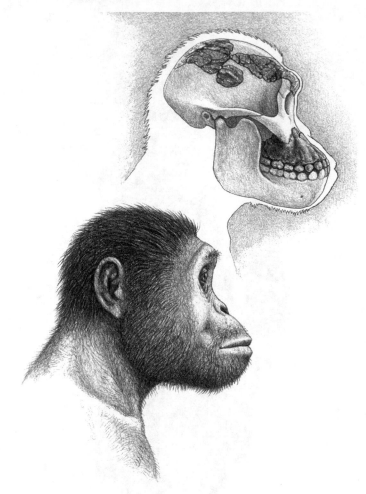

FIGURE 3.24

Life reconstruction of the head of *Australopithecus garhi*

The fragmentary cranial remains on which this reconstruction is based, from the Bouri Formation in the Middle Awash Valley, Ethiopia, clearly show that *Australopithecus garhi* had a primitive-looking, ape-like profile in side view. The maxilla was considerably prognathic and the nasal area very flat, so the outline of the face in side view was decidedly concave.

Awash Valley dated to even older deposits, around 5.8 to 5.2 Ma, also has been referred to *Ardipithecus ramidus* but given status as a separate subspecies, *A. r. kadabba*, on the basis of still more primitive dental characteristics. That would place it even closer to a common ancestor of humans and chimpanzees.

The third discovery comes from the Bouri Formation in the Middle Awash Valley in deposits dated to 2.5 Ma and consists of various cranial, dental, and postcranial elements scattered over quite an extensive area and evidently representing more than one individual. The cranial material of a single individual, BOU-VP-12/130, has been referred to yet another new species, *Australopithecus garhi* (figure 3.24), although it is not clear whether the postcranial elements represent the same taxon. The fourth find,

FIGURE 3.25

Life reconstruction of the head of *Kenyanthropus platyops*

The partial cranium on which the new species *Kenyanthropus platyops* is based is considerably crushed and damaged, but nonetheless is reasonably complete, giving a good indication of the general shape and proportions of the head. The face was very flat, thus resembling the East Turkana skull attributed to *Homo rudolfensis*, although this specimen, from the Nachukui Formation at Lomekwi, West Turkana, Kenya, not only is geologically older but also has many primitive features that prevent attributing it to the same species. However, it is not impossible that it is close to the direct ancestry of *H. rudolfensis*.

remains from the Nachukui Formation at Lomekwi, West Turkana, in Kenya, has been placed in a new genus and species: *Kenyanthropus platyops* (figure 3.25). The material dates to around 3.5 Ma and is considered different from that of any of the known australopithecine taxa, but possibly more similar to the material assigned to *Homo rudolfensis*, which might in this scheme also be referred to *Kenyanthropus*. The fifth is the so-called millennium ancestor, named because of its discovery in the year 2000, based on specimens from deposits dated to between 6 and 5.5 Ma at Lukeino, again in Kenya. The fossils are said to represent a bipedal hominin, despite retaining more primitive, ape-like dental features, and are referred to a new genus as *Orrorin tugenensis*. The claimed bipedalism at this date, its finders Martin Pickford and Brigitte

Senut argued, would make this an ancestor of modern humans and remove from the direct human lineage most, if not all, other hominins referred to *Australopithecus* and *Ardipithecus*. This argument has since been topped by the publication in 2002 of six specimens—a fairly complete cranium, a couple of mandible fragments, and some isolated teeth—from latest Miocene deposits, with dates of 7 to 6 Ma, at Toros-Menalla in the Djurab Desert of the Chad Basin. The material has been assigned to yet another new taxon, *Sahelanthropus tchadensis*. This hominin, too, is claimed to show an early development of upright, bipedal features in fossils that present a mosaic of primitive and more advanced characters.

It is clear that much more work has to be done in assessing these six sets of material. The *Orrorin tugenensis* specimens, for example, on which such a major argument is based, cannot even be shown to come from a single individual, let alone one species. The paleontological literature is full of what have later proved to be mistaken associations of taxa, producing what at first sight suggests a strange mixture of traits in a single species. The *Sahelanthropus tchadensis* material appears, according to the authors, to support the argument for an early development of bipedalism, but, again, the apparent mixture of traits must be examined and understood, and already there have been suggestions that it is no more than perhaps a primitive gorilla. The original *Ardipithecus ramidus* material still has not been fully described, to the frustration of many, making it difficult to see how the formal designation of newer material as a subspecies can be justified. In any event, the overall relationship of the gracile australopithecines to *Homo* is unresolved.

A very generalized scheme might see *Ardipithecus ramidus* or even its newly described "subspecies," *A. ramidus kadabba*, as a reasonable common ancestor, the earliest hominin after the split from the chimpanzees, with a plausible lineage then running through *Australopithecus anamensis* to *A. afarensis*. If *A. afarensis* is then considered primitive in comparison with the remaining hominin taxa, *A. africanus* and the robust forms in the genus *Paranthropus* can be regarded as its descendants. But what of the *Homo* lineage? If the material referred to *Homo habilis* and *H. rudolfensis* is removed from any closer association with the *Homo* lineage, it is difficult to link *A. afarensis* to *H. erectus*. The new species *Australopithecus garhi* has been claimed by its discoverers as a potential direct ancestor for the genus *Homo*. However, in their scheme, *Homo* includes the contentious taxa *H. habilis* and *H. rudolfensis* and provides a link for *A. garhi* because it shares megadont (large-toothed) characters with *H. rudolfensis*. Irrespective of opinions about the removal of *H. habilis* and *H. rudolfensis* from *Homo*, few commentators appear to agree with that view.

Overall, the confusion that appears to have been engendered by the more recent discoveries in Africa may simply be a reflection of our ignorance. It may be, as Bernard

Wood has argued, that later Miocene and Pliocene morphology in advanced apes was much more variable than previously assumed, and that the mosaic of traits now beginning to appear as we get a better record simply reflects that complexity.

Order Creodonta: Archaic Predators

The Creodonta contains two families of archaic, carnivorous animals: the Oxyaenidae and the Hyaenodontidae. Although a separate order from the modern Carnivora, the true carnivores, its members spanned the range of sizes and morphological types seen in the living carnivores and therefore may be regarded as their counterparts (figure 3.26). Creodonts differ from true carnivores in the modification of their cheek teeth for meat eating. In the Carnivora, the upper fourth premolar and the lower first molar form the carnassials, whereas in the creodonts they are formed by the upper first or second molar and the lower second or third molar. This difference suggests a separate evolutionary history and hence an order-level distinction; it is even possible that the creodonts themselves do not form a single natural group.

The Oxyaenidae, although also known from Asia, were distributed mainly in North America, and only the Hyaenodontidae are represented in Africa.

Family Hyaenodontidae

Members of the Hyaenodontidae first appeared in Africa during the late Eocene. Occurrences during the Oligocene were rare, but the family appears to have diversified greatly into around 10 species and was well represented by the early Miocene before becoming extinct by around 13 Ma. This late occurrence in Africa is matched only by that in Asia, since the hyaenodontids had become extinct in North America and Europe by the end of the Oligocene. Many are referred to the genera *Isohyaenodon*, *Metapterodon*, and *Hyaenodon*, but the family included the gigantic *Megistotherium osteothlastes* (plate 11) and is most widely represented by the species *Hyainailourus sulzeri* (figure 3.26). This was a very large animal with a head around twice the size of a modern lion skull that is found in deposits in northern, eastern, and southwestern Africa variously dating from around 22 to around 13 Ma. In an ecological sense, the hyaenodontids were therefore the forerunners of the true carnivores as the dominant meat-eating terrestrial animals in Africa.

Order Carnivora: True Carnivores

With around 250 living species distributed across virtually every habitat, the true carnivores, members of the Carnivora, make up one of the largest mammalian orders.

FIGURE 3.26
Size comparison of *Hyainailouros sulzeri*, *Amphicyon giganteus*, and *Crocuta crocuta*

Based on partial remains from Rusinga Island in Lake Victoria, Kenya, this reconstruction represents a specimen of moderate size, perhaps a female, of the gigantic species *Hyainailouros sulzeri* (*left*). Although the skull of this creodont was larger than that of any true carnivore, the overall proportions of this animal were like those of other hyaenodontid creodonts, with huge heads in relation to their bodies. The large bear-dog *Amphicyon giganteus* (*center*), which was sympatric with *Hyainailouros* at Arrisdrift in Namibia, would have been nearly as large as the latter, in spite of having a much smaller skull. Both species would have dwarfed the modern spotted hyena (*Crocuta crocuta*) (*right*), which is the largest carnivore in modern African ecosystems with a comparable combination of dental adaptations to eating meat and cracking bones. One can imagine that the disputes around a carcass in early Miocene times were rather spectacular with such huge scavengers involved.
Reconstructed shoulder height of *Hyainailouros sulzeri*: 1 m.

While the creodonts were the ruling predators of the Oligocene and into the earliest Miocene of Africa, the carnivores already had begun to disperse into the continent by the later stages of that time. Among the earliest to appear were the bear-like Amphicyonidae, the Felidae, and the Viverridae, but the Hyaenidae arrived shortly after. The Canidae, though, were relatively late arrivals, dispersing from their point of origin in North America much more recently and making it to Africa only in the late Miocene.

Family Hyaenidae: Hyenas

Around 70 living and fossil species of the Hyaenidae have been identified, but only four survive. The largest is the spotted hyena (*Crocuta crocuta*) (figure 3.27), at weights of up to 90 kg, followed by the brown hyena (*Parahyaena brunnea*), the striped hyena (*Hyaena hyaena*), and the aardwolf (*Proteles cristatus*). The status of the last has often been disputed in view of its aberrant features: while looking externally like a miniature striped hyena, it has a severely reduced dentition and subsists on a diet largely of

FIGURE 3.27
Spotted hyena

Modern hyenas are unmistakable because of their unique body proportions, with high shoulders and a sloping back, as shown by this spotted hyena (*Crocuta crocuta*) in the Masai Mara National Reserve, Kenya. Such proportions are regarded as adaptations that allow hyenas to carry large pieces of carcasses in their jaws and to travel enormous distances with their apparently tireless walk and cantering gallop. (Photograph by Mauricio Antón)

termites. Brown and striped hyenas may weigh up to 55 kg, but in general the brown tends to be larger and more robust, while the aardwolf reaches only 12 kg. Only the striped hyena now lives outside Africa, although the spotted hyena once was widely distributed in Europe and Asia.

Hyenas, in general, and the three larger ones, in particular, have received a very bad press, cast since the time of Victorian explorers in the role of mere scavengers, compared unfavorably with the supposedly noble big cats, and generally considered vermin to be shot on sight. Their habit of raiding graveyards and their employment as garbage disposers in some villages have hardly added to their reputation. In more recent years, careful field studies, stimulated by the excellent work of the naturalist Hans Kruuk on spotted hyenas during the 1970s, have shown hyenas to be much more interesting animals than their image would suggest, with often complex social arrangements. While they undoubtedly scavenge, they also hunt and may display highly cooperative hunting techniques among members of the clan. Indeed, Kruuk found that in the Ngorongoro Crater on the edge of the Serengeti Plain in Tanzania, the resident spotted hyenas were regularly killing their own prey, only to have up to 70 percent of it scavenged from them by lions. The feature that makes the larger hyenas such accomplished all-around carnivores is their impressive ability to consume bone and digest the organic fraction, seen most notably in the spotted hyena. This makes for the efficient consumption of carcasses, allowing hyenas to obtain a meal, especially in times of scarcity, from something of no value to another predator unable to smash and eat the bones. Their habit of collecting bones in and around the den makes them of particular interest to paleontologists, since it is clear that such behavior in the past has produced many of the bone assemblages found in caves and other former dens.

The oldest known hyenas are European specimens from early Miocene deposits of around 17 Ma, although the earliest African record is only a little later. The African record is patchy, however, and the next representatives are from a much later date, close to 12 Ma, with the appearance of *Protictitherium punicum* and *Lycyaena crusa-fonti* at Bled ed Douarah in Tunisia. The extant large hyenas have lineage records that go as far back as the mid to later Pliocene, while aardwolves, although perhaps different from the living form at the species level, are known from earliest Pleistocene deposits at the South African sites of Swartkrans, Kromdraai B, and Sterkfontein,

With only four living species from such a large number known in the fossil record, it might be assumed that the present range of variation of types is small in comparison with that seen over time. Studies by Lars Werdelin and Nikos Solounias have distinguished six adaptive types, or ecomorphs, among the living and fossil Hyaenidae, and shown that what we think of as the typical bone-cracking morphology is a fairly late development. The types are

1. Civet-like insectivore/omnivore
2. Mongoose-like insectivore/omnivore
3. Jackal- and wolf-like meat and bone eater
4. Cursorial (adapted to running) meat and bone eater, such as the members of the genus *Chasmaporthetes* (figure 3.28)
5. Transitional bone cracker
6. Fully developed bone cracker

The earliest African hyenas belong to types 1 (*Protictitherium*) and 2 (*Proteles* lineage). Upper Miocene faunas are dominated by types 3 (*Ictitherium* [figure 3.2; plate 12] and *Hyaenictitherium*) and 4 (*Lycyaena*). Type 3 (*Hyaenictitherium*) and primitive type 4 (*Hyaenictis*) hyenas continue to appear in the earliest Pliocene at Lange-baanweg, in the Cape Province of South Africa, but soon disappear. The largest hyena of all, the short-faced *Pachycrocuta* (figure 3.29), appears in the later Pliocene record of eastern and southern Africa and continued into the earlier Pleistocene, although it is better known in Eurasia where it existed during the whole of the early Pleistocene. This animal was substantially larger than the largest spotted hyenas, and may have weighed half as much again.

The greatest diversity of the African large-carnivore guild occurred around 4 to 1.5 Ma, when various extinct taxa overlapped for a considerable period with extant species. In Africa, as in Eurasia, type 3 hyenas had been replaced by dogs, first recorded at Langebaanweg. Indeed, a great transition in the structure of the Hyaenidae occurred at the end of the Miocene with the appearance of both the dogs and the large, bone-cracking type 6 species in the form of *Adcrocuta eximia*. Figure 3.30 shows *Ikelohyaena abronia*, a plausible ancestor of one or another of the smaller living species of bone crushers.

Families Percrocutidae and Stenoplesictidae: "False Hyenas"

Members of the Percrocutidae were formerly included among the hyenas on the basis of general similarities, such as enlarged, bone-cracking premolars and reduced number of molars. Studies of the deciduous, or milk, dentition now show convincingly that the morphology of those teeth differ too much between the percrocutids and the hyenas for them to be closely related to each other and that the similarities in the adult dentition are convergent adaptations to similar functions.

Some paleontologists suggest that the percrocutids should be placed in a different family, the Stenoplesictidae, which would include genera such as *Stenoplesictis*, *Africanictis*, and *Percrocuta* that are known from a variety of localities in northern, eastern, and southwestern Africa in a time span covering the Miocene. *Stenoplesictis* and *Africanictis* contain species of small size in relation to modern hyenas, rather like the small and more primitive insectivore/omnivore hyenas of the Miocene, although *Percrocuta* was a large animal with a well-developed, bone-cracking dentition paralleling that of some of the later large hyenas. This tendency to large size and bone-crushing teeth was particularly well developed in the species *Dinocrocuta algeriensis* from northern Africa.

Family Nimravidae: "Paleofelids"

The nimravids, broadly similar in many features of their anatomy to the cats, have at times been considered good members of the Felidae—hence the designation "paleofelids." Many of them differ sufficiently from the true cats to be placed in a separate family, the Nimravidae, and some researchers have even argued that they are in reality more closely related to the dogs. However, their morphological similarities to the true cats—with retractable claws, well-developed canine teeth, and cheek teeth

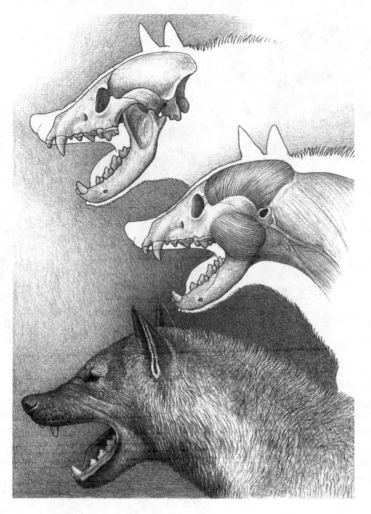

FIGURE 3.29

Sequential reconstruction of the head of *Pachycrocuta brevirostris*

This reconstruction of the head of *Pachycrocuta brevirostris* is based on a well-preserved cranium with mandible from the site of Kromdraai in the Sterkfontein Valley of South Africa. The skull and dentition show all the features typical of bone-cracking hyenas: stout build, with high zygomatic arches and a domed forehead; enlarged and robust premolars; and insertion areas for the masseter and temporalis that bespeak the enormous development of these muscles.

FIGURE 3.28

Life reconstruction of *Chasmaporthetes australis*, drawn to scale with *Crocuta crocuta*

This reconstruction from a partial skeleton of *Chasmaporthetes australis* (*top*), from the site of Lange-baanweg in the Cape Province of South Africa, clearly shows the size and proportions of this extinct hyena. It was considerably taller than any extant species, such as the spotted hyena (*Crocuta crocuta*) (*bottom*), but was a lightly built animal with a head relatively smaller than that of its modern relatives. The hind limbs were longer in proportion to the forelimbs, which gave the living animal a less sloping back. The teeth of *Chasmaporthetes* were less adapted to cracking bones, and more to chewing meat and skin, than those of living hyenas, all of which conjures up the image of a more active predator. Reconstructed shoulder height of *Chasmaporthetes australis*: 88 cm.

FIGURE 3.30

Life reconstruction of *Ikelohyaena abronia*

This reconstruction from a partial skeleton of *Ikelohyaena abronia*, from Langebaanweg in South Africa, shows an animal with body proportions that anticipate those of the modern hyenas. The hind limbs already were shorter than the forelimbs, although the difference was not as great as in the modern species, and the skull and dentition were those of a specialized bone cracker. However, *I. abronia* retained some primitive features, such as a relatively long tail and a clawed, reasonably developed first digit in the manus. Reconstructed shoulder height: 68 cm.

adapted for slicing meat—suggest similar lifestyles, although the nimravids showed an even greater tendency toward the development of a saber-toothed upper-canine morphology. Thus whereas only some of the true felids bore such teeth, all the nimravids were equipped with such fangs. In general terms, the nimravids of the Oligocene and earlier Miocene seem to have occupied the place taken over by true cats from the middle Miocene onward. Recent work by the paleontologist Jorge Morales and colleagues reinforces earlier suggestions that at least some of the nimravids, now placed in the subfamily Barbourofelinae, do fall closer to the Felidae.

An African presence for the nimravids is unclear. Two species have been identified: *Syrtosmilus syrtensis*, found in mid Miocene deposits of around 17 Ma in Libya, and *Sansanosmilus palmidens*, found at Bled ed Dourah in Tunisia and dated to around 14 to 10 Ma. However, Morales and colleagues have also placed both *Syrtosmilus* and *Sansanosmilus* in the Barbourofelinae. Debate over this matter undoubtedly will continue.

Family Felidae: Cats

The cats, or "neofelids," appear to have originated in Eurasia, where the earliest known is a small animal known as *Proailurus lemanensis*, found in French deposits of about 30 Ma. It was about 40 cm high at the shoulder and, although it probably looked perfectly cat-like, had a more primitive dentition than living cats, with more teeth in the jaw. The earliest African cat is a slightly larger animal of early Miocene age from the locality of Gebel Zelten in Libya. It has been variously named *Afrosmilus africanus*, *Metailurus africanus*, and *Pseudaelurus africanus*—the uncertainty reflecting the difficulty of assigning fragmentary material of primitive members of the Felidae to any one taxon. In addition, *Afrosmilus* has been suggested as a possible member of the Barbourofelinae.

Today the African felids are quite diverse, with nine from the total of 37 living species. They range in size from the lion (*Panthera leo*) (figure 3.31), in which large males may weigh in excess of 260 kg, through the leopard (*P. pardus*) (figure 3.32) and the cheetah (*Acinonyx jubatus*) (figure 3.33), whose largest populations are in Namibia, to the small black-footed cat (*Felis nigripes*), found in arid areas of southern Africa and weighing only 1 to 2 kg. They include the African wildcat (*F. lybica*) (figure 3.34), a plausible ancestor of the domestic cat known in large numbers from dynastic Egypt. The swamp cat (*F. chaus*) may be added to the list, although it is found only in the lower valley of the Nile. With the marked exception of the lion, most cats are variably solitary in their behavior, although the cheetah may live in bachelor groups of brothers or of subadults with their mother.

Lion prides may consist of up to a dozen adults with offspring. Females, often related, form the core and do most of the hunting and looking after the young on a broadly communal basis. Adult males, perhaps two or three and also often related to one another, take over prides from resident males and kill any cubs before fathering new offspring. Male offspring are driven out as they mature, but females may join the pride. The adult males defend the pride against seizure by other males that may roam in all-male groups in search of prides that they can appropriate.

All the living African felids are thoroughly cat-like in appearance and behavior. Most have long tails, sharp and retractable claws, large and pointed canine teeth, and cheek teeth that are largely adapted for eating meat and soft tissues. However, the smaller to medium-size serval (*Felis serval*) and caracal (*F. caracal*), in which males may weight up to 18 kg, are unusual in having shorter tails. Coat color and patterning are variable, and markings are clearly a widespread feature of the family. Even

FIGURE 3.33
Female cheetah with two cubs

The bond between cheetah (*Acinonyx jubatus*) siblings is very strong, and male cubs often remain together after they leave their mother, forming coalitions that can last for life. Although both cats are spotted, leopards and cheetahs differ greatly in appearance. The slender body and limbs of the cheetah betray its cursorial hunting style, and, given its lesser muscular strength, the cheetah cannot take prey as large as that brought down by the leopard. (Photograph by Mauricio Antón)

FIGURE 3.31
Male lion

There is considerable variation in the development and color of the mane in the male African lion (*Panthera leo krugeri*), and it is difficult to establish clear differences from the Asiatic lion (*P. leo persica*), which often is said to have a smaller mane. Another feature said to distinguish Asiatic from African lions is a marked fold in the skin along the belly, but some African lions, like this male in Etosha National Park, Namibia, show a similar fold. (Photograph by Mauricio Antón)

FIGURE 3.32
Female leopard

The leopard (*Panthera pardus*) is capable of taking very large prey relative to its own size. This female in Samburu National Park, Kenya, probably weighed some 40 kg, but it had killed a large impala ram that would weigh about 50 kg. Leopards have been reported to bring down much larger animals on occasion, thanks to the enormous strength of their well-muscled forelimbs and to the powerful bite of their canine teeth, which can strangle ungulates up to the size of an adult kudu. (Photograph by Mauricio Antón)

FIGURE 3.34
African wildcat

The African wildcat (*Felis libyca*) is the most likely candidate to be the ancestor of the domestic cat. It is a shy, secretive animal, and seeing one, like this individual in Chobe National Park, Botswana, is usually a matter of luck. (Photograph by Mauricio Antón)

lions, in which the adults are a fairly simple yellowish or grayish color, have young with spotted skins showing rosettes. Cheetahs, generally patterned with black spots on a yellow background, can have coats without any patterning or with the spots joined to give a marbled effect. They may also be largely black-coated, a condition known as melanism after the pigment melanin, which produces the darker hairs of the patterns. Melanism is not uncommon among leopards (the so-called black pan-thers) and servals, particularly in highland populations where it may indicate a rela-tively isolated gene pool. Even in these melanistic individuals, however, the basic pat-terning of the coat can still be seen when the light is right.

The large cats, in particular, generally capture prey using their claws, usually after a short stalk and chase, and kill by means of a neck bite or suffocation. Female lions frequently hunt cooperatively, although individuals can take fairly large prey, and pride males generally leave the hunting to the females (figure 3.35). Prideless males as well as pride males in more wooded environments hunt together, however, and take

FIGURE 3.35
Female lion with a giraffe

Bringing down an adult giraffe is a very risky business for lions, but prides in some parts of Africa have developed efficient techniques to take such dangerous prey. In several areas in southern Africa, giraffes become an important resource in the dry season, when other ungulates like zebra and wildebeest leave the territories of resident lions. This adult giraffe bull was hunted by a pride of lions nicknamed the Maome pride, which inhabited the Savuti Marsh in Chobe National Park, Botswana. (Photograph by Mauricio Antón)

large antelopes such as buffalo if they must. The leopard has its own specialization, often carrying into trees prey that may exceed its own body weight of up to 90 kg in order to keep it from the attention of other predators, particularly hyenas. The great strength that this requires is of double advantage to a cat that hunts alone and by stealth, usually in broken and wooded terrain where it can find both cover and refuge. The cheetah, with its long body and legs and lighter weight of up to 70 kg, has the most specialized hunting technique. It involves a high-speed chase, in more open country with patches of cover, of the smaller gazelles and a trip of the prey before dispatching it in the normal manner. Family groups may cooperate to some extent and tackle even fairly large prey, such as male impalas. Some of the smaller cat species hunt well in open terrain and may favor it.

Much of the fossil record of the Felidae is, however, dominated by the saber-toothed cats of the subfamily Machairodontinae. Before the mid Pliocene, the African felid guild consisted almost entirely of machairodonts, the saber-toothed and "false" saber-toothed cats of the genera *Dinofelis* and *Adelphailurus* (tribe Metailurini), *Machairodus* and *Homotherium* (tribe Homotheriini), and *Megantereon* (tribe Smilodontini). *Machairodus* appears to have been a late Miocene immigrant and *Homotherium* a true African genus, although the origins of *Dinofelis* are rather unclear, and *Megantereon* probably entered the continent during the early Pliocene. The sole known exceptions to this early pattern of machairodonts are two new species of conical-toothed cats from deposits dated to 17 Ma at the locality of Arrisdrift in Namibia, both placed in a new genus as *Diamantofelis ferox* and *D. minor*. The former is about the size of a cheetah; the latter, the size of a caracal.

The genus *Dinofelis*, members of which are often referred to as false saber-tooths because they lack the highly derived, flattened canines of the other taxa, is first recorded at Lothagam in Kenya, where it has been referred to a new species, although the extent to which it differs from later taxa is unclear. The material found at Langebaanweg, in the Cape Province of South Africa, is very similar to *Dinofelis barlowi*, a cat known from Transvaal deposits of late Pliocene and earliest Pleistocene age (figure 3.36). The latest is the more derived species, *D. piveteaui*, known from Kromdraai A, in the Sterkfontein Valley of South Africa, at close to 1.7 Ma (figure 3.37). All were around the size of a large modern jaguar (*Panthera onca*), and material from Langebaanweg associated with dental remains suggests a powerfully built animal (figure 3.38). *Dinofelis* is last recorded in eastern Africa in the Okote Member of the Koobi Fora Formation at East Turkana in Kenya and at Konso-Gardula in Ethiopia at around 1.5 Ma, while unpublished material listed as *D. piveteaui* is present at Kanam East in Kenya, but the date is uncertain.

The genus *Adelphailurus* is basically an American taxon based on a fragmentary skull and dentition from deposits of Hemphilian age at the Edson Quarry in Kansas. The remains suggest an animal the size of a puma. An African distribution is indicated by a specimen from the early Pliocene site of Langebaanweg. This specimen, a maxillary fragment from an animal also about the size of a puma or a small leopard, was originally identified as *Felis obscura*.

The genus *Homotherium* has been generally considered as a descendant in one form or another of the large Miocene machairodont genus *Machairodus* (figure 3.39). The latter, with a number of species, is not well represented in Africa, although a small sample has been identified at Langebaanweg in the southernmost tip of the continent. Recent reviews of material from the earlier deposits at Lothagam have complicated the picture by referring a large machairodont from there to a new genus and species, *Lokotunjailurus emageritus* (figure 3.40). This animal appears to combine features of both *Machairodus* and *Homotherium*, although it seems to differ too much from both to be accommodated in either.

Homotherium latidens, a larger animal rivaling male lions in size, also first appears

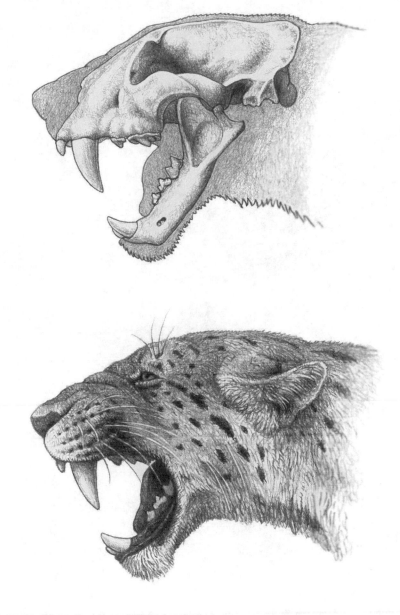

FIGURE 3.36

Life reconstruction of the head of *Dinofelis barlowI*

The Pliocene site of Bolt's Farm in South Africa has provided the best fossil known to date of *Dinofelis barlowi*. The overall proportions of the skull resemble those of a modern pantherine cat, but there are some slight, incipient "saber-tooth" features: the upper canines are a little longer than those of a pantherine of similar size, and they are more flattened in cross section, with anterior and posterior keels. The carnassials are relatively very large, and the mastoid process is somewhat more developed than that of modern big cats.

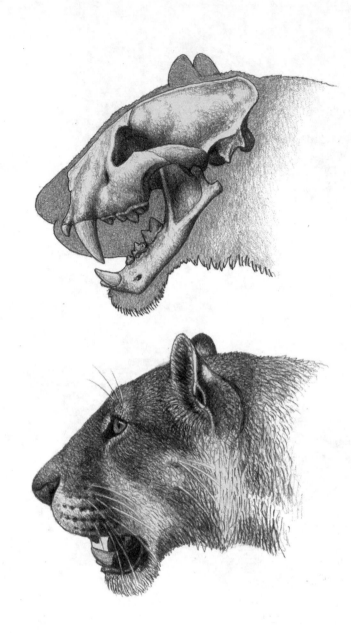

FIGURE 3.37

Life reconstruction of the head of *Dinofelis piveteaui*

The skull of *Dinofelis piveteaui* is best known thanks to a well-preserved specimen from the site of Kromdraai A, in the Sterkfontein Valley of South Africa, although more and more material is showing up at various localities in eastern Africa, including the Koobi Fora Formation at East Turkana, Kenya. This species is similar in body size to the earlier *D. barlowi*, but the skull is more specialized. The upper canines are more flattened (although not more high-crowned), the carnassials are proportionally larger, and the mastoid process is enlarged. The muzzle is considerably shorter, and, as a result, the diastema (the space between the canines and the first cheek teeth) is very small in both mandible and maxilla.

FIGURE 3.38

Comparison of *Dinofelis barlowi* and *Panthera pardus*

Based on cranial and postcranial remains from Bolt's Farm in South Africa, supplemented by information from other sites, this reconstruction shows similarities and differences between *Dinofelis barlowi* (*left*) and the modern leopard (*Panthera pardus*) (*right*). *D. barlowi* was considerably larger, with more robust forelimbs, somewhat shorter hind limbs, and a shorter tail. In the more derived saber-tooths, like *Megantereon* and *Homotherium*, the hind limbs were even shorter relative to the forelimbs. Reconstructed shoulder height of *Dinofelis barlowi*: 70 cm.

at Langebaanweg and is a relatively frequent find at sites in eastern Africa until its last appearance at West Turkana, in deposits dated to close to 1.5 Ma. In South Africa, it is rare in the Transvaal caves, but occurs at Makapansgat Limeworks, in Members 4 and 5 at Sterkfontein, and at Kromdraai A. *Homotherium* is remarkable for the relative length of its front limbs, giving it an apparently awkward posture, but it was a powerful animal whose limb length may have provided considerable leverage. Figure 3.39 shows a reconstruction of the head of the eastern African specimen referred to the species *Homotherium hadarensis*. Material from the lower Pleistocene locality of Incarcal in Spain shows evidence of cursorial adaptation in this genus; the claws are mostly small, while the bones of the forelimb are long and relatively slender.

Megantereon cultridens is often considered similar in size and overall build to a large jaguar, but was in fact highly variable in stature, with smaller animals more the size of a leopard and, by implication, rather sexually dimorphic (figure 3.41; plate 13). The cat is first clearly recorded in the South Turkwell locality in East Africa at around 3.5 Ma, perhaps in Member B of the Shungura Formation, and then fairly regularly in the later Shungura members until its last appearance there in Member G. The latest record in eastern Africa is once more in the Okote Member of the Koobi Fora Formation, dated close to 1.5 Ma. In South Africa, it is recorded from Kromdraai A,

FIGURE 3.39

Life reconstruction of the head of *Homotherium hadarensis*

This reconstruction of the head of *Homotherium hadarensis* is based on a relatively complete skull from Hadar in Ethiopia, but the upper canine and the mandible were restored on the basis of *Homotherium* fossils from other sites. The skull and dentition of this lion-size cat show all the traits of a specialized machairodont, including hypertrophied mastoids, a lowered glenoid, and a short zygoma. These and other features enabled the scimitar-toothed cat to open its jaws wide enough to clear the tips of the "sabers" and bite the throat of its prey with the combined force of the jaw-closing muscles and the head-depressing muscles of the neck, a process known as the canine shear-bite.

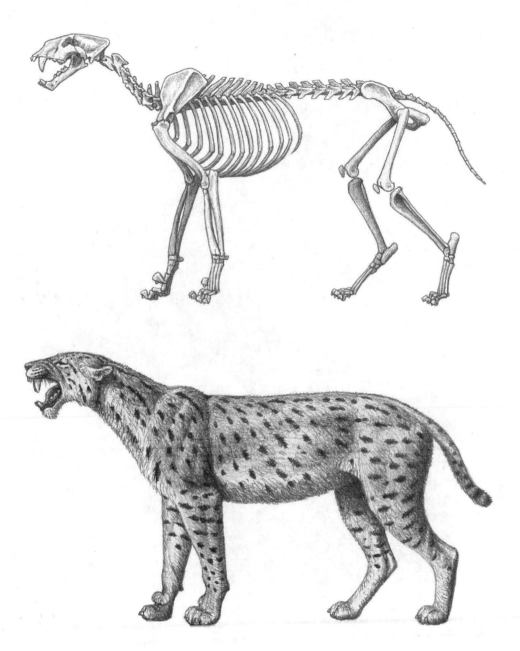

FIGURE 3.40

Life reconstruction of *Lokotunjailurus emageritus*

This reconstruction of *Lokotunjailurus emageritus* is based on an exceptionally complete skeleton from the site of Lothagam in Kenya, which served as a basis for the description of this new species of homotherine, or scimitar-toothed, cat. *Lokotunjailurus* was about as tall at the shoulders as a small lion but somewhat more gracile in build, with more elongated metapodia and a relatively smaller head. The claws were proportionally smaller except for that of the first digit of the forepaw, or dewclaw, which was enormous and would have been a visible feature in the living animal, despite being partly covered with skin and fur. Reconstructed shoulder height: 85 cm.

FIGURE 3.41

Sequential reconstruction of the head of *Megantereon cultridens*

A nearly complete skull and associated mandible of *Megantereon cultridens* from the Koobi Fora Formation at East Turkana, Kenya, provide a clear image of the cranial morphology of the African saber-toothed cat. No upper canines were found with this specimen, so they have been reconstructed based on material from South Africa. Like *Homotherium*, *Megantereon* was a very specialized machairodont (much more so than the species of *Dinofelis*), and it also killed prey using a canine shear-bite. There were differences in detail, however, especially in the shape of the upper canines, suggesting that there also were differences in the details of the biting method.

Basal length of skull: 21.4 cm.

Kromdraai B, Swartkrans, and Sterkfontein, although it is unknown at Makapansgat Limeworks.

Knowledge of the origins of the living large cats is strangely incomplete. There are no obvious likely ancestors in the fossil record, and paleontologists are left simply with a pattern of earliest records. The first appearances of the lion, leopard, and cheetah probably are in the Laetoli Beds at Laetoli in Tanzania, dated earlier than 3.46 Ma. Lions are rare in eastern African deposits, with the first evidence after Laetoli being in Member G of the Shungura Formation, at around 2.3 Ma, although there are several records of "large felinae" from the Usno Formation and from Members C, D, and F of Shungura, suggesting the presence of lions in present-day Ethiopia from perhaps 3 Ma. At East Turkana, they are recorded in only the Okote Member at around 1.6 to 1.4 Ma, but are known from Olduvai Gorge in Tanzania. In South Africa, the earliest appearance is in Member 4 at Sterkfontein, currently dated to around 2.8 to 2.4 Ma.

Family Viverridae: Genets, Civets, and Linsangs

There are arguments among taxonomists about the status of the taxa in the Viverridae. Some would include the mongooses as a subfamily, the Herpestinae, but we treat them as a separate family, the Herpestidae. All viverrids are most closely related to the hyenas and cats. With 72 species worldwide, the Viverridae is the most diverse family. There are nine African species of genet, two of civet, and two of linsang. Genets (figure 3.42) are largely arboreal, as is the palm civet (*Nandinia binotata*), while the African civet (*Civettictis civetta*) is terrestrial. Some viverrids are very widely distributed, although linsangs are restricted to western central Africa. Most are small animals, weighing no more than 2 to 3 kg, although the African civet, found in much of sub-Saharan Africa, can weigh up to 20 kg. The fossil history of the family is patchy, but some of the earlier specimens from Pliocene localities at Laetoli in Tanzania and Langebaanweg in South Africa have been referred to a single large species, *Viverra leakeyi* (figures 3.43 and 3.44). An even earlier viverrid, of the fossil genus *Kanuites*, comes from the Miocene site of Fort Ternan in Kenya (figure 3.45).

For many, the viverrids represent a good approximation of the primitive ancestral condition in the carnivores. However, one of the Madagascar species, the fossa (*Cryptoprocta ferox*), has a very derived appearance convergent on that of the cats, and it is extremely similar in the morphology of its postcranial skeleton to the earliest known cat (*Proailurus lemanensis*), found in France in deposits dated to around 30 Ma.

Family Herpestidae: Mongooses

There are 23 species of mongoose in Africa, found largely in the tropical, forested, and southern arid parts of the continent (figure 3.46). They are terrestrial animals, liv-

FIGURE 3.42
Common genet

Genets have many adaptations to arboreal locomotion, but they forage primarily on the ground, mostly at night. Although some researchers have stated that the genet is a plantigrade animal, when walking or trotting on the ground, it moves in a digitigrade gait, as seen clearly in this common genet (*Genetta genetta*), turning to a plantigrade locomotion when moving along the branches of trees. They are usually shy animals, but in some places they are less difficult to see, as in the camp in Samburu National Park, Kenya. (Photograph by Mauricio Antón)

ing on invertebrates and smaller vertebrates, and are perhaps most famously known for their ability to deal with poisonous snakes. Many live communally, and those inhabiting open territory, like the meerkats (*Suricata suricata*) of southern Africa, operate a sentry system to watch for such predators as raptorial birds. Members of the Herpestidae vary in size from a few hundred grams of body mass in the slender mongoose (*Herpestes sanguinea*), to 4 kg in the long-snouted mongoose (*H. naso*). As with the viverrids, the fossil history of the family is patchy, but it seems likely that the African species derive from a generalized ancestor similar to the living genus *Herpestes* that immigrated in the earlier Miocene.

Family Canidae: Dogs

The Canidae, with 35 living species worldwide, is divided into three living and fossil subfamilies—the Caninae, Borophaginae, and Hesperocyoninae—although only the Caninae is known from Africa. The continent has five living species of fox: red fox (*Vulpes vulpes*), Cape fox (*V. chama*), sand fox (*V. pallida*), Ruppell's fox (*V. ruppelli*), and fennec (*V. zerda*). In addition, Africa is home to the Ethiopian wolf (*Canis simen-*

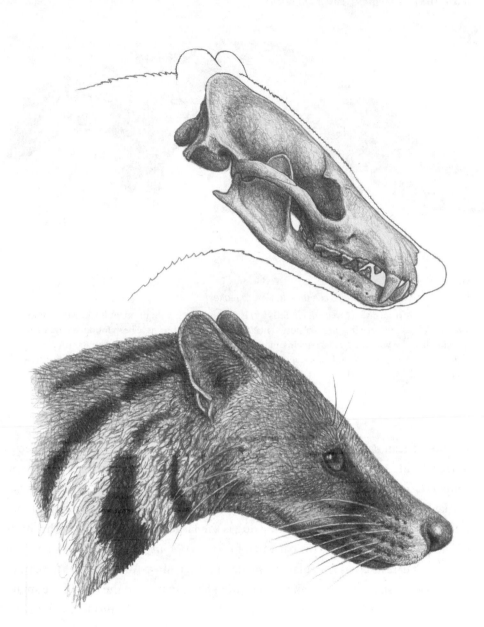

FIGURE 3.43

Life reconstruction of the head of *Viverra leakeyi*

This reconstruction of the head of *Viverra leakeyi* is based on a well-preserved skull from Langebaanweg in South Africa that, in general shape, is very similar to that of modern Asian ground civets of the genus *Viverra*, but corresponds to a much larger animal, about the size of a jackal. The dentition of *V. leakeyi* is somewhat more sectorial (that is, adapted to cutting meat) than that of modern members of the genus or the closely related African civet (*Civettictis civetta*). This feature suggests that *V. leakeyi* was more of an active predator of small and medium-size prey.

FIGURE 3.44

Size comparison of *Civettictis civetta* and *Viverra leakeyi*

The modern African civet (*Civettictis civetta*) (*left*) is drawn to scale with the giant fossil civet *Viverra leakeyi* (*right*). Only very fragmentary postcranial remains of *V. leakeyi* have been found, but they correspond well in size with the cranial remains, indicating that the living animal would be about 50 cm high at the shoulder and thus considerably taller than its modern relative or even a jackal. The coat pattern is reconstructed to resemble that of Asian civets of the genus *Viverra*.

sis), the African or Cape hunting dog (*Lycaon pictus*) (figure 3.47), and three species of jackal: side striped (*C. adustus*), black-backed (*C. mesomelas*) (figure 3.48), and golden (*C. aureus*). The foxes range in body weight from around 1.5 kg for the fennec to 8 kg for the red fox; the jackals are larger, at up to 16 kg; and the Ethiopian wolf may weigh up to 19 kg. The Cape hunting dog is the largest African canid, with a body weight of up to 36 kg, while the bat-eared fox (*Otocyon megalotis*) is only 5 kg but has enormous ears and has a diet largely made up of termites. All the rest of the canids are opportunistic predators, taking everything from insects, in the case of the smallest species, through larger invertebrates, small vertebrates, and fruits to medium-size antelopes, in the case of the pack-hunting wild dogs. The Ethiopian wolf, probably a relatively recent immigrant from Eurasia in view of its genetic similarity to the true wolf (*C. lupus*), has a very restricted range in the Ethiopian highlands, where it subsists mainly on rodents. Some of the foxes also have limited ranges, usually in more open areas of vegetation. The golden jackal is found in northern and eastern regions; the black-backed jackal, in northeastern and southern areas; and the side-striped jackal, over much of sub-Saharan Africa outside the central western forests, the Horn of Africa, and the southernmost region.

The fossil history of the family is poor in Africa and shows little, if any, presence

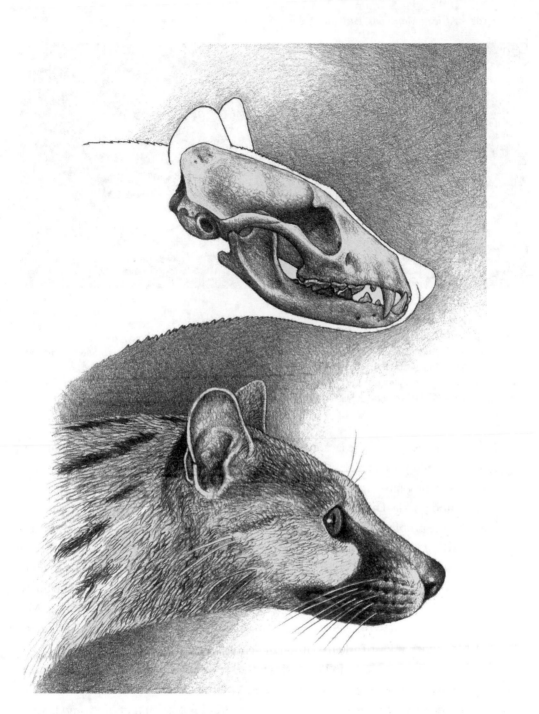

FIGURE 3.45

Life reconstruction of the head of *Kanuites*

A beautifully preserved skull of an adult viverrid of the genus *Kanuites* from the Miocene site of Fort Ternan in Kenya served as basis for this reconstruction. The adult mandible had to be reconstructed, since the site has yielded only juvenile mandibles, with their milk dentitions. While this carnivore has been classified as a mongoose in the past, the adult skull is most similar to that of a modern genet, although somewhat larger, and the external appearance of the living animal also would have been very genet-like.

FIGURE 3.46
Banded mongoose

Mongooses are better adapted for locomotion on the ground than are genets and most civets, and they walk on digitigrade feet with nonretractable claws. As seen in this banded mongoose (*Mungos mungo*) in Chobe National Park, Botswana, the claws in the forefeet are relatively large and the ears are reduced, features that are related to moderate burrowing abilities and are shared by all members of the family Herpestidae. (Photograph by Mauricio Antón)

for the larger species, the jackals and hunting dogs, before about 3 Ma. Foxes appear to have been present at Langebaanweg in South Africa, however, and a specimen placed in the genus *Canis* recently has been reported at South Turkwell in East Africa at around 3.5 Ma. Difficulties in separating the jackal species in fragmented samples add to the problems. The hunting dog is almost unknown in the fossil record, with the earliest clear evidence being that from Elandsfontein in the Cape Province of South Africa, where the deposits appear to be at least of middle Pleistocene age. There is a second possible occurrence in the deposits at Gladysvale in the Transvaal. Some material from the Kalochoro Member of West Turkana, with a date of 2.3 to 1.9 Ma, has been referred to the genus *Lycaon*, and there is mention of a mandible fragment of *Lycaon* size at East Turkana, although neither may be a correct identification. One other possible occurrence in East Africa is the skull from Olduvai originally described by Heinrich Pohle in 1928 as a new species: *Canis africanus*. The morphology of the upper and lower carnassials seems to agree more closely with that of *Lycaon* than of *Canis*, but the exact provenance of the specimen is unknown. One other oddity is the presence of a species of the Eurasian raccoon dog (*Nyctereutes terblanchei*), which usually is referred to the Caninae and has been identified at Kromdraai A, in the Sterkfontein Valley of South Africa, in deposits of around 1.6 Ma and at Ahl al Oughlam in Morocco, dated to 2.5 Ma (figure 3.49).

FIGURE 3.47

African hunting dog

The African hunting dog (*Lycaon pictus*) is often regarded as an animal of the open grasslands, but it is
perfectly at home in more wooded environments, such as the riverine landscapes of the Okavango Delta
in Botswana. Hunting dogs may even fare better in such habitats because the lower visibility helps to
hide their kills from hyenas and lions. (Photograph by Mauricio Antón)

FIGURE 3.48
Black-backed jackal
The black-backed jackal (*Canis mesomelas*), like this one in the Masai Mara National Reserve, Kenya, is an elegant canid. These jackals are nocturnal when confronted by humans, but can be easily seen during the day in protected areas, when they forage alone or in pairs for small animals and carrion. (Photograph by Mauricio Antón)

An interesting point is the appearance and rise to prominence of the Canidae in comparison with the changes in the Hyaenidae. Earlier, pre-Pliocene hyenas were much more generalized in their appearance than modern hyenas and, by implication, ecological roles, with many of them having a more dog-like appearance. This correlated pattern of change between hyenas and dogs is also seen in Eurasia at the same period.

Family Mustelidae: Honey Badger, Weasels, Polecat, and Otters
With 67 living species worldwide, the Mustelidae is a highly diverse family, although only 11 species occur in Africa, usually placed in two subfamilies.

SUBFAMILY MUSTELINAE: HONEY BADGER, WEASELS, AND POLECAT
Among the Mustelinae, the honey badger, or ratel (*Mellivora capensis*), similar in size and build to a European badger at up to 16 kg, is known from most parts of sub-

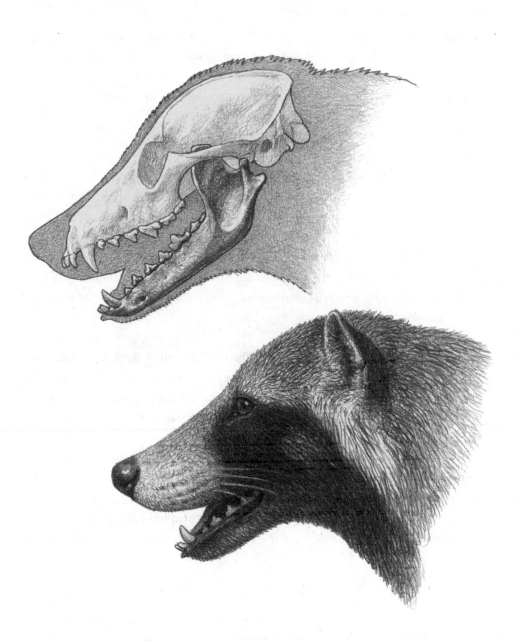

FIGURE 3.49

Life reconstruction of the head of *Nyctereutes terblanchei*

The mandible of this Eurasian raccoon dog is based on almost complete fossils assigned to *Nyctereutes terblanchei* from Kromdraai A, in the Sterkfontein Valley of South Africa, while the cranium is reconstructed on the basis of related species of the genus. The coat pattern follows that of the raccoon dog (*N. procyonides*), the only extant species of the genus.

Saharan Africa. The weasels are much smaller. The white-napped weasel (*Poecilogale albinucha*), weighing about 350 g, is found in much of the southern half of the continent. The Libyan striped weasel (*Poecilictis libyca*), about twice that weight, is restricted to the north, and the striped polecat (*Ictonyx striatus*), larger at about 1.4 kg, is found in sub-Saharan areas outside the Congo Basin.

SUBFAMILY LUTRINAE: OTTERS
The Lutrinae contains the otters, which consist of four species, two of which are clawless—the Cape clawless otter (*Aonyx capensis*), found in most southern and western regions)—or have reduced claws—the Congo clawless otter (*A. congica*), found in the Congo Basin. The common otter (*Lutra lutra*) is confined to the northwestern coastal area, while the spotted-necked otter (*L. maculicollis*) has a distribution similar to that of the Cape clawless otter. The spotted-necked otter is much the smallest, with a body weight of up to 6.5 kg, while the common otter may weigh up to 20 kg or more and the two clawless species reach close to 30 kg.

Much work clearly remains to be done on the fossil record of the Mustelidae. Fossil mustelids are recorded in Africa from early Miocene times onward, with the otter-like *Kenyalutra* known from around 20 Ma. A large, extinct genus of otter, *Enhydriodon*, dates from the late Miocene and Pliocene of eastern Africa and the Pliocene of Namibia, and what has been described as a hypercarnivorous, medium-size species, *Namibictis senuti*, is recorded from Arrisdrift in Namibia. The Cape clawless otter is identified from Pleistocene deposits at Swartkrans in the Transvaal. A large mustelid from late Miocene deposits at Lothagam in Kenya, placed in the new genus *Ekorus*, appears to have been of leopard size, suggesting that members of the family may have gone some way toward evolving a cat-like morphology and, by implication, a cat-like lifestyle (figure 3.50).

Family Ursidae: Bears
Of the two extant subfamilies of the Ursidae, one is recorded in Africa, as is an extinct group of ursids.

SUBFAMILY URSINAE
Tribe Ursini
Bears as we think of them—brown bears, black bears, polar bears, and other members of the tribe Ursini—are not generally thought to have entered Africa, although the Ursidae seems to have originated in Eurasia and to have colonized the New World successfully on several occasions. However, a small number of teeth from the late

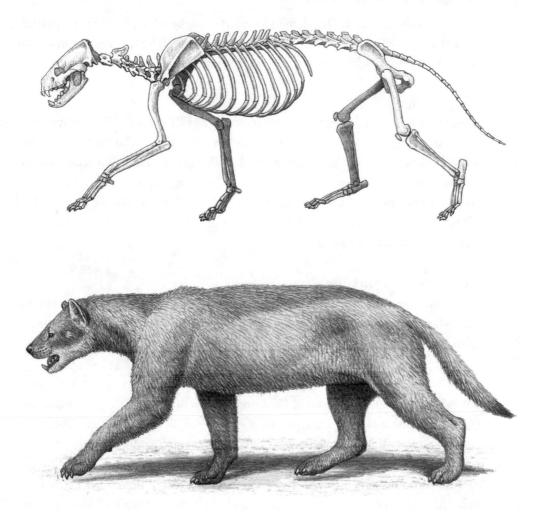

FIGURE 3.50

Life reconstruction of *Ekorus ekakeran*

The giant mustelid *Ekorus ekakeran*, described recently from the site of Lothagam in Kenya, not only was much larger than the biggest living terrestrial mustelid from Africa, the honey badger (*Mellivora capensis*), but also differed from it in its functional anatomy. The bones of the forelimb, in particular, show that the muscles associated with digging, so well developed in the honey badger, were less powerful and its posture was upright, as in a cat or hyena, rather than crouching, as in the badgers. The resulting image is of a large, relatively fast predator that would have actively pursued medium-size prey. Reconstructed shoulder height: 61 cm.

Pliocene locality of Ahl al Oughlam in Morocco have been identified by the paleontologist Denis Geraads as similar to those of the Etruscan bear (*Ursus etruscus*), an extinct European species, and more work clearly needs to be done on this matter.

Tribe Ursavini

The genera *Indarctos* and *Agriotherium*, members of the extinct tribe Ursavini, are recorded respectively from Sahabi in Libya and Langebaanweg in South Africa at close to the Miocene–Pliocene boundary, and it has always been clear that such material should turn up somewhere between these two localities (figure 3.51). *Agriotherium* has now been reported from the hominin site of Aramis in the Middle Awash Valley of Ethiopia, with a date of around 4.4 Ma, so the biogeographic connection now appears to have been established.

SUBFAMILY HEMICYONINEA: "HALF DOGS"

The extinct Hemicyoninae, or "half dogs," does seem to have entered Africa in the early Miocene interchange around 20 Ma, marked by the presence of the genus *Hemicyon* in eastern Africa.

Family Amphicyonidae: Bear-dogs

Not to be confused with the ursid "half dogs" of the subfamily Hemicyoninae, the Amphicyonidae was a separate and now extinct family of superficially bear-like carnivores of medium to large body size that first appeared in the Eocene. Although most commonly found in North America and Eurasia, the amphicyonids are also known from Africa. The name bear-dog stems from the Amphicyonidae having been included with the Canidae, the true dogs, and most taxonomists consider the bears and dogs, along with the mustelids and raccoons, to be closely related to one another within the order Carnivora.

The genus *Cynelos* is recorded from eastern Africa at around 20 to 18 Ma, while *Amphicyon giganteus* (figure 3.26) and *Ysengrinia ginsburgi* have been identified at Arrisdrift in Namibia at around 17 Ma. *Agnotherium* appears later, at around 14 to 11 Ma. Two species are recorded from late Miocene deposits at Lothagam in Kenya, perhaps the youngest known members of the family. The bear-dogs presumably were rather generalized carnivores, with a greater reliance than living bears on meat in their diet, and perhaps also specialist scavengers, using their size to good advantage in appropriating kills.

FIGURE 3.51

Life reconstruction of *Agriotherium africanus*

This reconstruction of *Agriotherium africanus*, based on cranial and postcranial material from the site of Langebaanweg in South Africa, shows the great similarities between *Agriotherium* and modern bears. However, the limbs of the extinct ursid were relatively a little longer than those of a modern brown bear, and this feature, coupled with slight differences in the dentition and skull, has led some specialists to regard *Agriotherium* as a more active predator.

Reconstructed shoulder height: 1.3 m.

Order Embrithopoda: Arsinoitheres

Members of the Embrithopoda have no known ancestors or descendants, although they probably were most closely related to the Proboscidea and the Sirenia and, as such, are good candidates for inclusion in the Afrotheria. Until recently, the order was clearly recorded only in the Oligocene deposits of the Fayum Depression in Egypt, from which the single species *Arsinoitherium zitteli* is described, although material is now known from Ethiopia. The animal, shown in a life reconstruction in figure 3.52, resembled a rhinoceros in size and general appearance, although it had a pair of large bony horn cores attached to the skull, each of which had a smaller knob set at its rear border. These cores perhaps were covered in a sheath of horn made of keratinized skin, as is seen in the true horns of the Bovidae, and differed from the compressed fibrous structures lacking any bony core that make up the so-called horns of living rhinos. The skeleton also suggests a lifestyle very different from that of rhinos, perhaps with a preference for aquatic conditions. Material from Turkey, Romania, and even Mongolia has been linked with the Egyptian species, but is much smaller and more primitive.

Order Proboscidea: Trunked Animals

The Proboscidea today is restricted to two species of elephant: the Indian (*Elephas maximus*) and the African (*Loxodonta africana*) (figure 3.53), which are the largest living land animals. Once widely distributed throughout Africa, elephants are now found only in restricted areas of the continent, although within those regions local populations may be considerable. Ironically, the very size and strength of the animal, which render adults virtually invulnerable to predators (figure 3.54)—a fully grown male can weigh over 6000 kg—bring it into conflict with humans, since its ability to get through all but the strongest barriers to raid crops makes it a pest. The value of its ivory has, of course, added to pressures on the population. Unlike the Indian elephant, the African species has been little exploited as a domestic animal, although it can be tamed and is used for excursions into the swamps of the Okavango Delta in Botswana and in Garamba National Park in the Republic of Congo. The Carthaginian general Hannibal Barca employed African elephants in his invasion of Italy in 218 B.C.E., although how he got them to his starting point in Spain is not entirely clear.

The herds are well organized matriarchally, so there always tends to be a congregation of a dozen or so animals, females and young, in all but the denser areas of woodland. Males of about 15 years leave the herd and may form bachelor groups. A

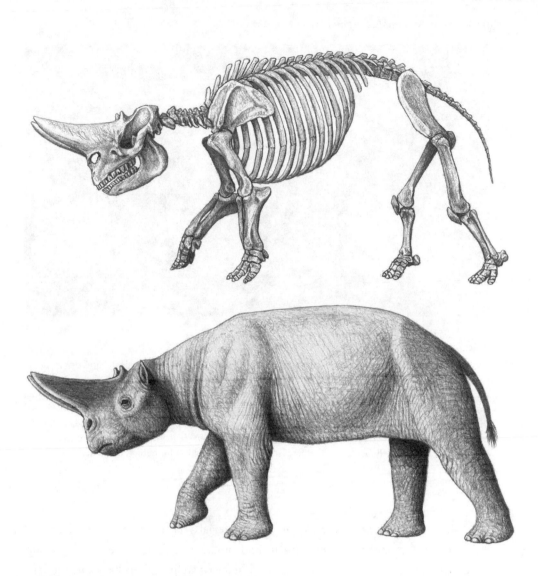

FIGURE 3.52

Life reconstruction of *Arsinoitherium zitteli*

The massive horn-like appendages in the snout of the large *Arsinoitherium zitteli*, described on the basis of remains from the Fayum Depression in Egypt, give it a rhinoceros-like appearance to a modern observer, but the similarities are in fact superficial. The "horns" of *Arsinoitherium* were structures of solid bone projecting from the nasal area of the skull, unlike the horns of rhinoceroses, which are made of hair-like tissue and have a bony base only on the skull. The postcranial skeleton was also very different from that of rhinos, which, although heavy and robust, are actually descended from cursorial ancestors and retain many adaptations for running in their limb bones. The limbs of arsinoitheres were columnar and ended in five-toed feet, generally resembling the limbs of elephants more than those of rhinos. Reconstructed shoulder height: 1.84 m.

FIGURE 3.53
African elephant

African elephants (*Loxodonta africana*) with impressively long tusks, like this female in Samburu National Park, Kenya, are becoming an increasingly rare sight in the wild owing to pressure from human hunters. Large tusks are important for display as signs of status and strength in mature individuals within elephant society, as weapons for fighting and defense, and as tools for food acquisition, and the fossil record shows that long-tusked individuals were common in many species of extinct proboscideans. In some modern populations, the pressure is so great that individuals genetically unable to grow tusks are selected because ivory poachers do not kill them, so entire herds have no tuskers at all. (Photograph by Mauricio Antón)

single large elephant can eat 300 kg of plant material a day and may live to the age of 60. Elephants consume a range of plants and can be very destructive of trees, which they either push over to obtain the higher foliage or strip of their bark during drier periods. Such damage can make it difficult to maintain breeding populations in smaller reserves. In dry areas and during droughts, elephants dig for water and are uniquely able to obtain any that collects in the bottom of holes, using a trunk that can hold up to a dozen liters. The close social structure provides a safe environment for the young, which are liable to be taken by lions up to the age of around four years, after which they become too large. But by the age of two years or so, the inquisitive juveniles tend to wander, and it is this age group, between two and four, that attracts most of the predator attention.

Proboscidean remains are large, make good fossils in the right conditions, and are easily seen in the field and recovered. But extraction is laborious and transportation difficult, so the postcranial bones are too often left at the site. The teeth are particularly characteristic, and enormous numbers are known. Yet despite such copious

FIGURE 3.54

Elephant guarding a water hole from a lion

Adult elephants are virtually immune to predation, and powerful enough to intimidate all other animals in the vicinity of water holes. Even lions, like this adult female from the Savuti Marsh in Chobe National Park, Botswana, have to wait patiently until elephants drink their fill. (Photograph by Mauricio Antón)

material, the study of relationships has produced numerous complex and often mutually contradictory schemes to explain the patterns of ancestry and descent. The origins of the order have long been thought to lie in Africa, where early Eocene deposits have produced *Moeritherium*, a small and perhaps generally hippo-like animal of around 300 kg body weight, with dispersal to Eurasia taking place in the early Miocene. This view is strengthened by the results of biomolecular analyses, which place proboscideans firmly in the Afrotheria.

Earlier members of the Proboscidea are divided into several families.

Family Elephantidae: Elephants

The Elephantidae, or true elephants, include the living Indian and African species and the mammoth. The fossil record of this family goes back to later Miocene times around 8 Ma, with particularly important material coming from the Pliocene of eastern Africa. The Indian genus, *Elephas*, originated in Africa and is well represented in the fossil record of the eastern part of the continent. The teeth are high crowned and reflect patterns of increasing adaptation to abrasive vegetation, particularly seen in successive populations of the species *Elephas recki* during the later Pliocene (figures 3.55–3.57). The earliest members of the modern African genus, *Loxodonta*, are known from late Miocene deposits in the Lukeino Formation, while mid Pliocene deposits at Kanapoi, also in Kenya, have produced the species *Loxodonta adaurora*. The extensive fossil proboscidean diversity in Africa included the ancestor of the well-known woolly mammoth (*Mammuthus primigenius*), featured in European ice age cave art. Dispersal of the mammoth lineage from Africa occurred at around 3 to 2.5 Ma.

Family Mammutidae: Mastodons

Members of the Mammutidae have very different teeth from the Elephantidae, lower crowned with more rounded cusps, a feature from which their name derives (Greek, *mastos* [nipple] and *odont* [tooth]). Mastodons should not be confused with the similar-sounding mammoths, which are true elephants. This family can be traced back to around 22 Ma in Africa.

Family Gomphotheriidae: Pig-toothed Proboscideans

The Gomphotheriidae, with their complex cheek teeth, is widely known, and early forms of this family are thought to have given rise to the true elephants of the Elephantidae. The gomphotheres are particularly well represented by the genus *Anancus*, an animal marked by very straight tusks (Greek, *anancus* [without curve]). The species *Anancus kenyensis* is known from the late Miocene of eastern Africa in the Lukeino Formation and in the Middle Awash Valley of Ethiopia; *A. petrocchi*, from

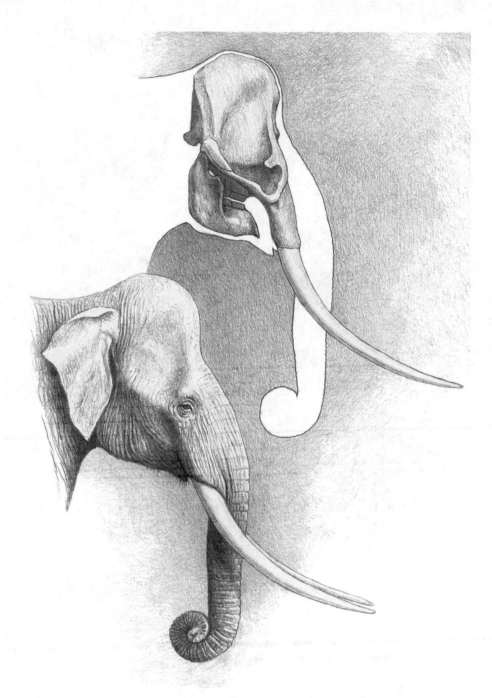

FIGURE 3.55

Life reconstruction of the head of *Elephas recki*

Complete skulls of the extinct *Elephas recki* have been found in the sites of the Omo River in Ethiopia and East Turkana in Kenya. They show the distinctively short and high head reminiscent of the modern Asian elephant (*Elephas maximus*), but with a forehead that was much more exaggeratedly domed. Such proportions are completely different from those of the head of the extant African elephant.

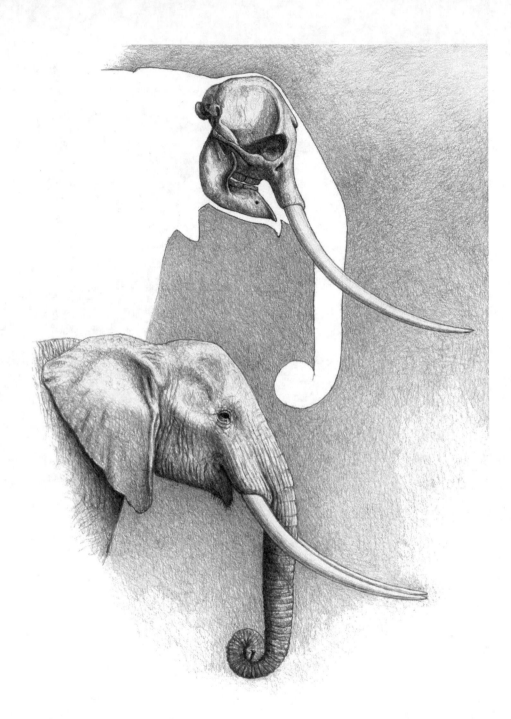

FIGURE 3.56

Head of African elephant

Compared with *Elephas recki*, the African elephant (*Loxodonta africana*) has a low skull, with a forehead that curves gently and shows hardly any sign of doming. A further important difference, not obvious from a general view of the skull, is the lesser complexity of the ridges of the molar teeth in the extant species, which indicates that it is less adapted than its fossil relative for eating abrasive fodder.

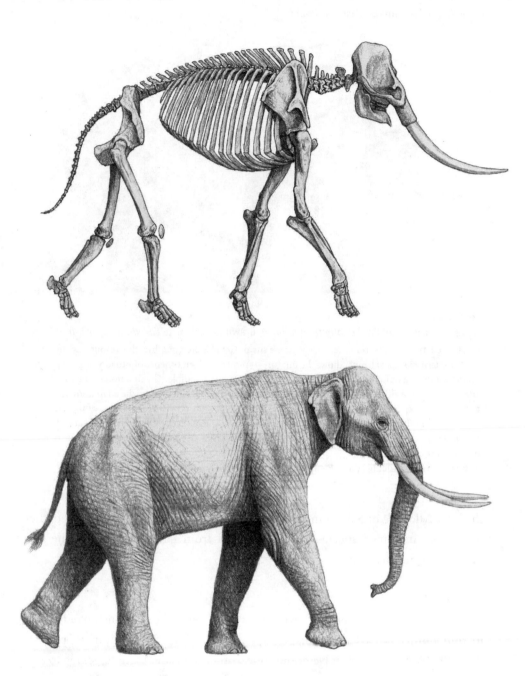

FIGURE 3.57

Life reconstruction of *Elephas recki*

A spectacular, nearly complete skeleton found in the Koobi Fora Formation of East Turkana, Kenya, served as a basis for this reconstruction of *Elephas recki*. The skeleton corresponds to an especially large individual, nearly 4.5 m high at the shoulder, and thus taller than the largest extant elephant bulls. The body proportions were unlike those of modern elephants, with a disproportionately small head and elongated forelimbs that gave the animal a somewhat sloping back. The tusks appear relatively small for the elephant's size, and the skull, with its inflated forehead and low-placed orbits, would contribute to the unique appearance of *E. recki*, which could never be mistaken for its living African relative.

FIGURE 3.58

Life reconstruction of *Prodeinotherium hobleyi*, drawn to scale with *Loxodonta africana*

Based on partial remains from the early Miocene site of Gebel Zelten in Libya, this reconstruction of *Prodeinotherium hobleyi* (*front*) shows the mixture, typical of all deinotheres, of broadly elephantine body proportions and a unique head with downturned tusks in the mandible. The function of the tusks remains a mystery, although they clearly were put to some use, as evidenced by the adaptations in the skull and neck vertebrae for strong and efficient vertical, especially downward, motions of the head. Although deinotheres of later age were large proboscideans, the modern African elephant (*Loxodonta africana*) would have dwarfed the early form from Gebel Zelten, especially a male as large as the one shown in this drawing.

Reconstructed shoulder height of *Prodeinotherium hobleyi*: 2.2 m.

the later Miocene of Sahabi in Libya; and *A. osiris*, from the Pliocene of northern Africa. An unnamed species is also recorded from Langebaanweg in the Cape Province of South Africa.

Family Deinotheriidae

Members of the Deinotheriidae have high crested teeth. The relationship of this family to the order Proboscidea is disputed, however. While some authorities place the deinotheres in the suborder Deinotheriodea, others consider them to belong to a distinct order, although a sister group to the Proboscidea. They are well known in the African fossil record, especially in the Pliocene of eastern areas with the species *Deinotherium bozasi*, elephant-like in size and structure and with a nasal region indicative of a trunk. Deinotheres are distinguished from other proboscideans by the absence of upper tusks and the presence of distinctive, downwardly curved tusks in the lower jaw and by the elongated shape of the skull. An earlier genus, *Prodeino-therium*, is known from Gebel Zelten in Libya (figure 3.58).

FIGURE 3.59

Life reconstruction of *Stegotetrabelodon syrticus*

The partial cranium and jaws on the basis of which *Stegotetrabelodon syrticus* was described come from the site of Sahabi in Libya, and are among the most spectacular proboscidean fossils known. In addition to having a pair of well-developed, straight upper tusks, *Stegotetrabelodon* had an elongated mandibular symphysis, from which a pair of long lower tusks projected forward and downward. The marked downward inclination of the symphysis and the low position of the occipital condyles suggest that the animal normally would have carried its head rather elevated.

Reconstructed shoulder height: about 3 m.

Proboscidean tusks are simply modified and elongated incisors. Many of the earlier proboscideans, such as the late Miocene gomphothere genera *Stegotetrabelodon* (figure 3.59) and *Gomphotherium*, had tusks in both the upper and lower jaws or only in the lower jaw, as in *Deinotherium*, whatever its status. The lower tusks often developed in a seemingly bizarre manner among the Gomphotheriidae to form spade- or shovel-like features, as in the genus *Platybelodon*, although this characteristic is not known from Africa.

Order Perissodactyla: Odd-toed Ungulates

Modern representatives of the Perissodactyla—hoofed ungulates—are the rhinos, horses, and tapirs. Members of this order are considered "odd-toed" because the central axis of the foot passes through the third or central digit and the weight-bearing digits are often reduced to an odd number. There are three in the rhinos and one in the horses (including asses and zebras), whose familiar single hoof often is shod in domestic animals for use on hard surfaces. Compared with the ruminant artiodactyls, the perissodactyls have a much simpler and more straightforward digestive system, one that is suitable for processing relatively poor, fibrous fodder fairly quickly and in bulk.

In the past, the Perissodactyla were considerably more diverse than they are today, especially the rhinos and horses, and they included some rather bizarre forms.

Family Chalicotheriidae: Claw-hoofed Ungulates

Members of the Chalicotheriidae, together with the rhinos, were probably the first of the perissodactyls to enter Africa in the wave of movements that followed definite contact with Eurasia after 20 Ma. Two groups are known on the continent. The chalicotheriines, in particular, presented an odd appearance to modern eyes: a horse-like head, elongated front legs, and claws rather than hooves, although those of the front feet were turned inward. Overall, the reconstructed stance in many ways resembles that of a gorilla (figure 4.3). The claws have been interpreted as adapted for digging, but the dentition is not heavily specialized and lacks evidence of an abrasive component in the diet, so the use of the long arms to attain vegetation for browse is perhaps implied. Members of the second group, the schizotheriines, were more horse-like in general proportions, but had claws on all four feet. The family is long-lived in the fossil record, and the schizotheriine species *Ancylotherium hennigi* is known from deposits as late as 2 Ma in eastern Africa. Chalicothere footprints are recorded from the site of Laetoli in Tanzania (figure 3.60).

Family Rhinocerotidae: Rhinoceroses

The Rhinocerotidae is represented by five species, two of which are found today in Africa: the white rhinoceros (*Ceratotherium simum*) and the black (*Diceros bicornis*), although the color difference may be difficult to discern (figure 3.61). Indeed, the term "white" has been said to derive from the Afrikaans word for "wide," used to describe the greater breadth of the mouth in this species. Both are classified as endangered, and the white rhino almost became extinct in the early twentieth century. With a weight of up to 2300 kg, the white rhino ranks as the third largest African mammal after the elephant and the hippo; at up to 1500 kg, the giraffe outweighs the black rhino. Both species obviously are free from natural predators. The white rhino is a purely grassland animal, and its long, low-slung head equips it very well for grazing. The black rhino carries its head much higher and browses on vegetation typically found on the edges of wooded areas. Both species lead essentially solitary lives, although the white rhino has been described as semisocial, but territoriality appears to be flexible, especially in the white rhino.

FIGURE 3.60

Life reconstruction of *Ancylotherlum hennigi*

A couple of unique footprints from the site of Laetoli in Tanzania correspond to a tridactyl, clawed ungulate about the size of a horse, and a detailed analysis of these prints has shown that they fit well with the anatomy of the feet of *Ancylotherium*, a chalicotherid genus known from the Miocene to Pleistocene of Africa and Eurasia. *Ancylotherium* walked on the tips of its metapodia, and the large claws usually were retracted so only the tips touched the ground. Fossils of the African species *Ancylotherium hennigi* are actually recorded at Laetoli, confirming the attribution of the footprints, but the remains are fragmentary, so this reconstruction is based mainly on the better known anatomy of the European species, *Ancylotherium pentelecicum*.

Reconstructed shoulder height 1.5 m.

Human activity, including poaching for the horn, has wreaked havoc on the rhinos' former distributions, but the fossil history of the African rhinos can certainly be traced back to the late Miocene, when they are known from a number of localities around the present-day Mediterranean. The two living species probably can be seen in Africa from at least mid Pliocene times and, in this sense, are two of the most successful mammals. The rhinos provide an object lesson in the misleading picture of

diversity in any group that may come from ignoring the fossil record. With the two African and three Asian species in great danger of extinction in the wild, we may easily overlook the fact that more than 16 rhino genera have been identified in the fossil record of Africa and Europe alone over the past 20 Ma (figure 3.62).

Family Equidae: Horses

The zebras typify Africa in the minds of many, but they are relatively very late arrivals on the plains of the continent. There are three living species (figure 3.63). The common zebra (*Equus burchelli*) is now found in eastern and southern Africa. The mountain zebra (*E. zebra*) lives in the arid areas of southernmost and southwestern Africa. Males of both species may weigh up to 320 kg. Grevy's zebra (*E. grevyi*) inhabits the northeastern arid zone of northern Kenya and Ethiopia. It is a larger animal, with males weighing up to 400 kg. The zebras are joined by the wild ass (*E. africanus*), which itself may weigh almost 300 kg and is now found in small populations along the southern borders of the Red Sea.

Zebra stripes may function as part of a mate-recognition system, as outlined in chapter 1, although it has also been suggested that they serve to confuse predators, help foals to recognize their mothers, or stimulate grooming. A large herd of zebras running in various directions when startled by a predator, or what they perceive as a possible threat from one, certainly makes a confusing picture. Even the wild ass has vestigial stripes on its legs, and such patterns can appear in domestic animals and the Asian wild, or Przewalski's, horse. Adult Grevy's zebras tend to live in mixed social groups, although females with young may congregate, but adult common and mountain zebras form "harems" of perhaps half a dozen females and young under the control of a stallion. Common zebras are the frequent prey of lions and spotted hyenas, with the latter forming hunting groups of up to 20 or so animals in order deal with

---

FIGURE 3.61

White and black rhinoceroses

The white, or square-lipped, rhinoceros (*Ceratotherium simum*) (*top*), such as this mother and calf in Ben Albert's Park, South Africa, is the second largest land mammal of Africa, weighing up to 2300 kg and having a shoulder height of 1.8 m. Several features of its anatomy are adaptations for grazing, including the shape of the upper lip and the large, elongated head, which is carried low on the neck and allows the muzzle to reach comfortably for the short grasses. The black, or hook-lipped, rhinoceros (*Diceros bicornis*) (*bottom*) is a browser, and its hook-like upper lip works like a "finger" to handle twigs and leaves, while its relatively short head is carried high on the neck. With a shoulder height of 1.6 m and a weight up to 1100 kg, the black rhino is smaller than its grazing cousin and even lighter than the giraffe, like the one in the background at Etosha National Park, Namibia. But in spite of this size difference, adult black rhinos are immune to predation (unlike giraffes) and can even relax and take a nap in broad daylight, ignoring the possible presence of lions. (Photographs by Alan Turner [*top*] and Mauricio Antón [*bottom*])

FIGURE 3.62

Comparison of *Dicerorhinus leakeyi* and *Brachipotherium heinzelini*

The presence of partial, articulated skeletons of these two rhinoceros species at the site of Rusinga Island in Lake Victoria, Kenya, allows us to appreciate the considerable differences in body proportions. Like other early members of the *Dicerorhinus–Stephanorhinus* group, *Dicerorhinus leakeyi* (*left*) was a large, tall rhinoceros with relatively long limbs, while *Brachipotherium heinzelini* (*right*) was a typical teleoceratine with short, robust legs and slightly hippo-like proportions, which are suggestive of semiaquatic habits. Reconstructed shoulder height of *Dicerorhinus leakeyi*: 1.6 m.

the protective aggression of the stallions and bring down the fast-moving females. Lions may similarly cooperate with other pride members in hunting zebras.

The origins of the Equidae possibly lie in North America during the Eocene, and, indeed, much of the evolution of the family also took place on that continent. Horse evolution is often portrayed as a classic story of change from a small, three-toed animal with low-crowned teeth to the long-legged, single-toed, high-crowned, grass-eating species of today. As with many popular portrayals of evolution, that is true but it is not the whole truth; the evolutionary pattern is not a single, ladder-like ascent but a bush, from which all but the long-legged, single-toed, high-crowned, grass-eating species have been pruned.

The earliest African equids were still three-toed animals, found in late Miocene

FIGURE 3.63

The three living zebra species

Although the easiest way to tell apart the extant species of zebras is to look at their different stripe patterns, these are not the only differences. Grevy's zebra (*Equus grevyi*) (*top*), photographed in Samburu National Park, Kenya, is taller and larger than the others and has a large, long head with big rounded ears. The mountain zebra (*E. zebra*) (*center*), seen at Lion and Rhino Game Reserve, South Africa, is similar in size and build to the common zebra (*E. burchelli*) (*bottom*), photographed in Etosha National Park, Namibia, but it displays a characteristic dewlap in the throat that is not present in the other species. (Photographs by Mauricio Antón)

deposits of around 10 Ma. They generally have been referred to a single genus, *Hipparion*, although current work by Raymond Bernor suggests that the relationships and taxonomy of the hipparionines are much more complex, although as yet not fully established. Two later dispersions of hipparionines into eastern Africa from Eurasia have been proposed, beginning at around 10 Ma with *Hippotherium primigenium* and followed at around 8 Ma by ancestors of the *Eurygnathohippus* lineage (figure 3.64). By the later Miocene, cursorial, open-country forms are known in addition to more robust species likely to have favored more wooded habitats (figures 3.65 and 3.66).

What are often referred to as the true horses, single-toed members of the genus *Equus*, again evolved in North America and spread across the Bering Land Bridge to Siberia, across Eurasia, and into Africa around 2.5 Ma. They therefore arrived in Africa during the major climate changes during the Pliocene and flourished in the newly opened vegetation. The various species of hipparionines, while hypsodont, had lower crowned teeth than the horses, but for some time around and after the true horses appeared, the hipparionines seem to have made an effort to adapt to the changing conditions by increasing the crown heights of their cheek teeth. One later species, *Eurygnathohippus cornelianus*, seems to have combined this heightened crown with further extreme adaptations of its eating apparatus that included an elongated snout and altered incisors to enable it to gain better access to short grass. Such animals were evidently successful in this evolutionary strategy, but eventually they had gone extinct by the mid Pleistocene.

Order Artiodactyla: Even-toed Ungulates

Africa has the most species-rich fauna of hoofed animals seen anywhere in the world, and the bulk of this diversity is made up by the Artiodactyla. These ungulates, themselves extremely diverse with almost 80 living genera in 10 families, are represented by pigs, peccaries, hippos, camels, chevrotains, deer, giraffes, antelopes, cattle, sheep, and goats. Members of this order are considered "even-toed" because the central axis of the foot passes between digits three and four, and the weight usually is borne on these two digits. Many of the artiodactyl families have an extensively modified alimentary system with a complex series of stomachs to allow a lengthy digestion process. The most developed of these animals—chevrotains, deer, giraffes, and antelopes—are grouped in the suborder Ruminantia, although the chevrotains do not have the fully elaborated system of four stomachs. The camelids have three effective stomachs, probably reflecting a common ancestry with the ruminants. The pigs, peccaries, and hippos are classified as the suiform artiodactyls in the suborder Suiformes—that is, pig-like animals—with peccaries and hippos having independ-

FIGURE 3.64

Life reconstruction of the head of *Eurygnathohippus cornelianus*

A nearly complete skull of a subadult *Eurygnathohippus cornelianus*, found at East Turkana in Kenya, reveals the very horse-like proportions of the head, with a long muzzle, large masseter muscles, and eyes set far back in the skull.

FIGURE 3.65

Comparison of *Eurygnathohippus turkanense* and *E. feibeli*

Cranial, dental, and postcranial remains from the site of Lothagam in Kenya reveal the presence of a large species of hipparionine, *Eurygnathohippus turkanense*, and a smaller one, *E. feibeli*. The limbs of the smaller horse were gracile and elongated, so while the difference in shoulder height was not great, the difference in body mass would be readily evident in the living animals, shown side by side in this reconstruction.

Reconstructed shoulder height of *Eurygnathohippus turkanense*: 1.2 m.

FIGURE 3.66

Life reconstruction of hipparionine and foal, compared with common zebra and colt

A remarkable sequence of hipparionine tracks from Laetoli in Tanzania has been interpreted as a foal crossing in front of its mother (*top*), a kind of behavior common in modern equids, as seen in this common zebra mare and colt in Etosha National Park, Namibia (*bottom*). The Laetoli hipparionines were using a gait called running walk, a lateral-sequence walk that provides considerable stability when moving on slippery surfaces, such as the wet volcanic ash that eventually would preserve the footprints as fossils. Even so, the mare could not help a slight slip when the colt crossed in front of her, which forced her to "brake," resting the lateral digits of her right hind foot and leaving a tridactyl print. Most other footprints show the mark of a single hoof, as would correspond to the normal locomotion of hipparionine horses. (Photograph by Mauricio Antón)

Reconstructed shoulder height of adult hipparionine: 1 m.

FIGURE 3.67

Warthog

The male warthog (*Phacochoerus aethiopicus*), like this one in Etosha National Park, Namibia (*top*), displays impressive tusks and very developed warts on his face. The female, like this one in Pilanesberg National Park, South Africa (*bottom*), has much less conspicuous tusks and warts and, in general, is smaller and more lightly built. This female is in the "kneeling" posture (the animal actually rests on the wrists of its forelimbs), which warthogs often adopt in order to reach short grass and tubers in spite of their relatively long legs and short neck. (Photographs by Mauricio Antón)

ently evolved an enlarged fore stomach. Several long-extinct families are also included in the order, but the only ones of concern here are the Anthracotheriidae and Climacoceratidae.

Family Suidae: Pigs

Today in Africa, the Suidae is chiefly represented by four species. The red river hog (*Potamochoerus porcus*) weighs up to 115 kg; the bushpig (*P. larvatus*) and the warthog (*Phacochoerus aethiopicus*) (figure 3.67), up to 150 kg; and the forest hog (*Hylochoerus meinertzhageni*), the largest, up to 280 kg. All have sub-Saharan distributions, with the red river hog inhabiting wet areas in the rain forest; the bushpig favoring thickets, forest, and bush; the warthog preferring lightly wooded or open terrain and grassland, where it may live in loose groups of related individuals and is a frequent prey of most of the larger predators; and the forest hog living in more restricted sections of denser forest in northern central Africa. A fifth species, the European wild boar (*Sus scrofa*), is found along coastal areas of northern Africa.

Various schemes have been proposed to accommodate the living pigs in subfamilies, with the warthog placed in the Phacochoerinae and the three other African species in the Suinae, or true pigs. Fossil pigs are known from Oligocene deposits in Eurasia, but first appear in Africa in deposits of around 17.5 Ma at Arrisdrift in Namibia and at sites such as Rusinga Island in Lake Victoria, Kenya, at around the same time. During the Miocene and Pliocene, the pigs were very diverse, and several taxonomies involve placing species in different genera and creating yet more subfamilies, including the early to mid Miocene African Kubanochoerinae (figure 3.68), Listriodontinae, and Hyotheriinae. Listriodonts were rather odd pigs, with the more evolved members of the subfamily having perhaps a greater amount of leafy vegetation in their diets, while hyotherins were small and generalized in their morphology and perhaps ancestral to the later pigs. The Tetraconodontinae replaced all these groups toward the end of the Miocene, after 10 Ma. These were pigs with thickened dental enamel, most characterized by species of the genus *Nyanzachoerus*, and seemingly immigrants at that point from Eurasia (figure 3.69; plate 14).

The current general consensus would see a linked group of the genera *Nyanzachoerus* and *Notochoerus* existing in the later Miocene and Pliocene, each genus with several species and *Notochoerus* perhaps arising from the species *Nyanzachoerus jaegeri*. The *Nyanzachoerus–Notochoerus* lineage had gone extinct by around 1.8 Ma with the disappearance of its last species, *Notochoerus euilus* and *N. scotti*, leaving no living relatives. Members of the genus *Notochoerus* were giants among pigs, the adults weighing up to 450 kg (figure 3.70). A second group, classified as the genus *Metridiochoerus*, is known from the Pliocene and earliest Pleistocene and has been regarded as the likely ancestor of the warthog. This genus underwent a major evolutionary

FIGURE 3.68

Life reconstruction of *Kubanochoerus massai*

A remarkably complete skull and various postcranial bones of *Kubanochoerus massai* were found at the site of Gebel Zelten in Libya, permitting a reliable reconstruction of the appearance and body proportions of this large suid. Like that of Eurasian members of the genus *Kubanochoerus*, the Gebel Zelten skull displays a couple of small conical appendages above the orbits, but it differs from the Eurasian specimens in lacking the large "horn" in the center of the forehead. Some specialists regard this difference as an example of sexual dimorphism, with the Libyan skull belonging to a female. Species of *Kubanochoerus* were very large, long-limbed pigs, exceeding in mass and height the largest extant members of the family. Reconstructed shoulder height: 97 cm.

turnover just after 2 Ma with the extinction of *Metridiochoerus andrewsi* and the appearance of *M. modestus*, *M. hopwoodi*, and *M. compactus*. Large size was again evident, with later representatives of *M. andrewsi* weighing up to 160 kg. Yet another group of pigs was represented in Pliocene and earliest Pleistocene deposits by species of the genus *Kolpochoerus* (figure 3.71), suggested as a plausible ancestor of the forest hog and bushpig.

The adults of these large pig species probably were too big and too aggressive to have been taken by any of the predators, but the young of all pigs are likely to have been an attractive prey, particularly before their tusks developed and especially because they presumably were left in a den while the adults foraged. Pigs have high reproduction rates in comparison with other ungulates, and a constant supply of piglets would have been a valuable resource for any predator.

FIGURE 3.69

Life reconstruction of the head of *Nyanzachoerus syrticus*

Well-preserved skulls of *Nyanzachoerus syrticus* from Lothagam in Kenya show that this species displayed strong sexual dimorphism, with the males being larger and having more ornamented skulls than the females. This reconstruction shows a male, which bore conspicuous bony ornaments on the muzzle and widely flaring cheekbones. In life, these ornaments would have been augmented by an extra thickness of skin, making them even more spectacular. The tusks, however, were of only moderate size. This type of cranial design is not very different from that of the modern forest hog (*Hylochoerus meinertzhageni*), if rather more extreme, suggesting a broadly similar fighting style for the males of the species.

FIGURE 3.70

Size comparison of *Phacochoerus aethiopicus, Metridiochoerus andrewsi,* and *Notochoerus euilus*

The typical height at the shoulder of the male warthog (*Phacochoerus aethiopicus*) (*left*) is between 65 and 85 cm, while the reconstructed shoulder height of *Notochoerus euilus* (*right*), based on a partial skeleton from the Koobi Fora Formation at East Turkana, Kenya, would be around 1.2 m. *Metridiochoerus andrewsi* (*center*) is known from excellent cranial remains at Koobi Fora and elsewhere, but unfortunately only fragments of its postcranial skeleton have been found, so its body proportions and size are inferred from skull size and the body proportions of extant suids.

Family Tayassuidae: Peccaries

The Tayassuidae, a family of pig-like artiodactyls, is today confined to New World. However, fossil peccaries are known from Europe and Asia and have been identified at the late Miocene locality of Langebaanweg in the Cape Province of South Africa, with the species *Cainochoerus africanus* (figure 3.72), and in deposits dated to between 16 and 12 Ma in eastern Africa. Although details are scarce, *C. africanus* has also been recorded at Lothagam in Kenya, although current opinions suggest that the attribution to the Tayassuidae is mistaken and that the material really belongs to a peccary-like true pig of the Suidae.

Family Anthracotheriidae: Pig- or Hippo-like Ungulates

Members of the Anthracotheriidae have a long fossil history and are known in Africa from Oligocene, Miocene, and Pliocene deposits. The sizes of species varied from small dog-like up to hippo-like, and the general appearance could be described as a mixture of pig and hippo morphology, with short, stout limbs; a broadly pig-like skull; and squared molars. Some anthracotheres are reported from aquatic deposits,

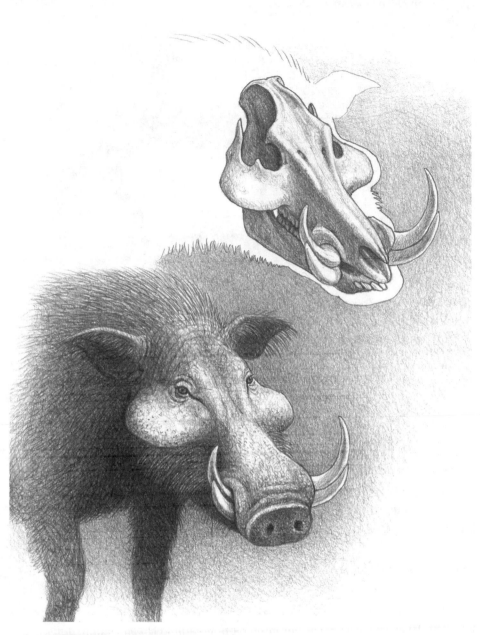

FIGURE 3.71

Life reconstruction of the head of *Kolpochoerus limnetes*

A beautiful, complete skull of a male *Kolpochoerus limnetes* from the Koobi Fora Formation at East Turkana, Kenya, served as a basis for this reconstruction, which shows the combination of massive zygomatic knobs with impressive recurved tusks that made this one of the most spectacular suids ever. The extant forest hog (*Hylochoerus meinertzhageni*) probably is closely related to *Kolpochoerus*, and it displays similar traits (large size, massive zygomatic knobs, and large tusks), but to a more moderate degree.

FIGURE 3.72
Life reconstruction of *Cainochoerus africanus*

The site of Langebaanweg in the Cape Province of South Africa has yielded cranial and postcranial remains of this small pig, once considered to be a peccary but now believed to be a diminutive member of the family Suidae. The limb bones of *Cainochoerus* are small and gracile, and their articulations have prominent ridges that restrict motion to the sagittal plane, indicating a cursorial adaptation more marked than in any other member of the pig family.

with numerous specimens found in the Fayum Depression in Egypt (figure 3.73). The genus *Merycopotamus*, known from Miocene and Pliocene deposits in Africa, is particularly hippo-like in morphology, with expanded flanges at the rear of the lower jaw and incisors and canines that appear very large in relation to the other teeth. The large *Merycopotamus petrocchii* is the most commonly found taxon at the site of Sahabi in Libya.

Family Hippopotamidae: Hippopotamuses

The Hippopotamidae is represented by two living species in Africa. The pygmy hippopotamus (*Hexaprotodon liberiensis*), of the forest and coastal plains of western Africa, weighs up to 280 kg. The common hippopotamus (*Hippopotamus amphibius*), at weights of up to 3000 kg, is the well-known species of sub-Saharan Africa (figure 3.74). The earliest hippos are known from Miocene deposits of 20 Ma in Kenya, with the appropriately named *Kenyapotamus*, and during the Pleistocene they emigrated from Africa, eventually becoming widely distributed in northern and central Europe during interglacial periods. Hippos were once common in sub-Saharan Africa wherever suitable watercourses existed.

Hippos may congregate in large groups, but social status must be constantly advertised and maintained by complex behavior patterns. Males are much larger than

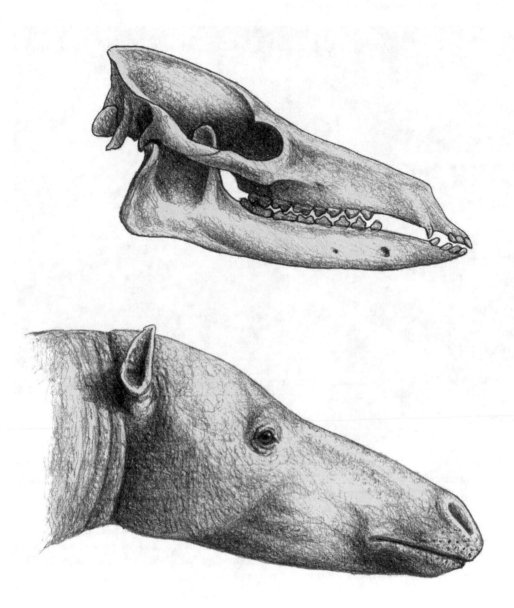

FIGURE 3.73

Life reconstruction of the head of *Bothriogenys fraasi*

The early anthracothere *Bothriogenys fraasi*, remains of which come from the Fayum Depression in Egypt, displays the typical cranial morphology of the family, with a low skull, a long facial region, and a complete dentition. In life, the head would have looked vaguely pig-like, and the whole animal rather like a cross between a pig and a hippopotamus.

FIGURE 3.74
Hippopotamus
During such yawning displays, the jaws of the hippopotamus (*Hippopotamus amphibius*) open to well over 100 degrees, and the anatomy of the skull and mandible shows profound adaptations for such enormous gapes. Such adaptations are also found in fossil hippo species, indicating that they had similar display behavior. (Photograph by Mauricio Antón)

females, with massive jaws and thick necks, and contests for mating rights involve much pushing and slashing with the jaws and front teeth and may be vicious. Such attacks on small boats can produce fatalities. At night, hippos leave the water to graze on grass in the vicinity, producing well-marked trails that often make access to the water easier for many other animals. Their size, together with their largely aquatic lifestyle, keeps hippos largely free from predation.

Until well into the Pliocene, the dominant genus in Africa was *Hexaprotodon*, with numerous species identified, including at least one dwarf form (figure 3.75). Hexaprotodont hippos were characterized by a more gracile skeleton, with narrower

FIGURE 3.75

Comparison of *Hexaprotodon aethiopicus* and *Hippopotamus amphibius*

A partial skeleton of *Hexaprotodon aethiopicus* (*left*) from West Turkana in Kenya allows a reconstruction of the overall appearance of this fossil pygmy hippopotamus. It was considerably smaller than the male common hippo (*Hippopotamus amphibius*) (*right*), but the difference would be less striking if it were compared with some small females of the extant species. In overall morphology, the postcranial bones are rather similar in both species, but the limb bones of the fossil are more slender, the lateral digits of the feet are more reduced, and the articulations are tighter, limiting lateral rotation of the feet. These are adaptations for a more efficient locomotion on dry ground and evidence of a less completely aquatic habit. However, the orbits are considerably elevated over the level of the skull roof, indicating a more advanced adaptation to life in the water than was present in earlier, more primitive species of the genus like *Hexaprotodon harvardi*.

Reconstructed shoulder height of *Hexaprotodon aethiopicus*: 1.1 m.

feet and limbs that had less rotational mobility, and seem to have been better adapted for movement on land than members of *Hippopotamus* (figure 3.76). In contrast, such features as the more elevated orbits and the extended frontal sinuses and facial region in *Hippopotamus* indicate a greater reliance on an aquatic environment and lifestyle. It seems likely that increasing aridity during the late Pliocene threatened the probable preferred habitat of the hexaprotodonts, the more widely distributed grass pastures and woodlands or thickets, and increased the advantage of exploiting the remaining bodies of water. Indeed, the later hexaprotodont hippos of the species *Hexaprotodon karumensis* display more elevated orbits than earlier species (figure 3.77), indicative of a more aquatic lifestyle, while retaining the terrestrial adaptations of the postcranial skeleton. Figure 5.7 and plate 15 show similar features in the head of the earlier *Hippopotamus gorgops*.

Family Camelidae: Camels

The Camelidae originated in North America in the late Eocene and was widespread and quite diverse there during the Miocene and into the Pleistocene—during which

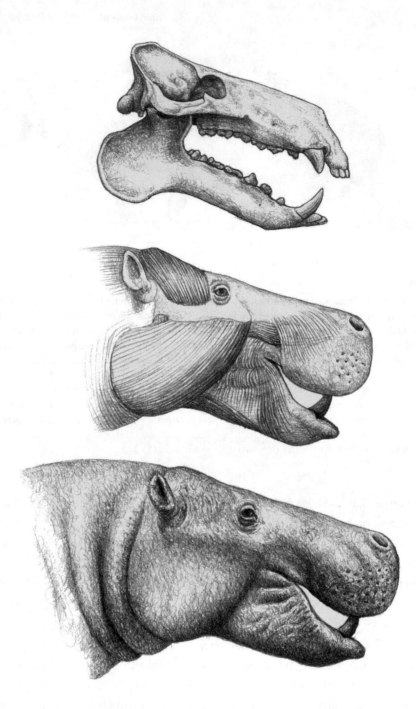

FIGURE 3.76

Sequential reconstruction of the head of *Hexaprotodon harvardi*

Beautiful cranial remains of *Hexaprotodon harvardi* have been found at Lothagam in Kenya, and they clearly show the primitive, although unmistakably hippopotamus-like, morphology of the head of this early species. The clearest difference for a modern observer would be the position of the orbits, which in the fossil hippo are almost level with the roof of the skull, instead of clearly raised or "periscopic," as in modern hippos.

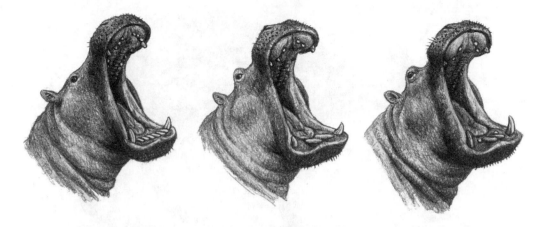

FIGURE 3.77

Comparison of the heads and dentition of *Hexaprotodon harvardi*, *Hexaprotodon karumensis*, and *Hippopotamus amphibius*

The number of upper and lower incisors traditionally has been one of the main features used for classifying hippopotamuses at the generic and specific level. The primitive, hexaprotodont (six anterior teeth) condition is well exemplified in the Lothagam species, *Hexaprotodon harvardi* (*left*), with six incisors in the maxilla and six in the mandible. The modern hippo (*Hippopotamus amphibius*) (*right*) has only four upper and four lower incisors, but the fossil *Hexaprotodon karumensis* (*center*), in spite of being classified as a hexaprotodont hippo, had only four upper and two lower incisors. In fact, the evolution from four to two lower incisors in the *H. karumensis* lineage took place within a few hundred thousand years, as seen in the different levels of the Koobi Fora Formation at East Turkana, Kenya. This shows how the trend to reduce the number of incisors can evolve independently in different lineages.

they probably spread to South America—going extinct in North America only at the end of the last glaciation. Members of this family are also well known in the fossil record of Asia. Camels appear briefly in the fossil history of Africa, being recorded in deposits of Pliocene age in the eastern part of the continent and even managing to get as far south as Malawi, where they are found in the Chiwondo Beds. But the animals now found in North Africa were introduced by humans.

Family Tragulidae: Chevrotains

The small, solitary water chevrotain (*Hyemoschus aquaticus*), the only member of the Tragulidae found in Africa, will be unfamiliar to many. Chevrotains are primitive ruminants that were widespread during the Oligocene. The water chevrotain weighs no more than 10 to 12 kg and looks rather like a small, short-legged deer without antlers. It lives today only in the lowland forests of central Africa, but members of the Miocene genus *Dorcatherium* were more widespread, from Arrisdrift in Namibia to a variety of localities in eastern Africa (figure 3.78). It is not recorded at Lothagam in Kenya, and therefore appears to have gone extinct by around 7 Ma.

FIGURE 3.78
Life reconstruction of *Dorcatherium chappuisi*

The fragmentary fossil remains of *Dorcatherium chappuisi* from Fort Ternan in Kenya indicate that this animal was little different from the better known members of the genus that lived in Europe in the Miocene and that, in turn, are almost identical to the extant African chevrotain (*Hyemoschus aquaticus*). Some specialists even suggest that both genera should be united into a single taxon. It appears that staying in riverine habitats and having an unspecialized anatomy and a varied diet have allowed chevrotains to remain virtually unchanged for about 20 Ma.
Reconstructed shoulder height: 35 cm.

Family Cervidae: Deer

The only member of the Cervidae that appears to be native to Africa is the so-called Barbary stag, a variety of red deer (*Cervus elaphus*) now confined to the Medjerda Mountains of the Tunisia–Algeria border area. It is a fairly large animal, with males weighing up to 225 kg and carrying antlers, which they shed annually. The Barbary stag is known from the fossil record of the past 1 Ma or so, and was then distributed across northwestern Africa from Tunisia to Morocco.

Family Climacoceratidae: Giraffoides

The Climacoceratidae and the Giraffidae usually have been grouped together in the superfamily Giraffoidea, although the giraffoides' precise relationship with the true giraffes remains to be established. The climacoceratids were medium-size to large ungulates with relatively elongated limbs and necks and variously developed protuberant structures on the skull. These "horns" differ from the ossicones of giraffes in being outgrowths from the frontal bones, and it is really the form of the lower canine teeth that suggests a link between the families. Most discussions have been based on fragmentary remains of *Climacoceras gentryi* from Fort Ternan in Kenya, but extensive material recovered from early and mid Miocene deposits at Elisabethfeld and Arrisdrift in Namibia in recent years may permit a clearer understanding of the evolutionary history of the climacoceratids. Three more species—*Sperrgebietomeryx wardi*, *Propalaeoryx austroafricanus*, and *Orangemeryx hendeyi*—are now known from those deposits (figure 3.79).

Family Giraffidae: Giraffes and Okapi

The true giraffes are relatively recent newcomers to the African fauna, first appearing in late Miocene deposits following dispersion from Eurasia. These members of the Giraffidae have an efficient, ruminant digestive system and were considerably more widespread and diverse in the relatively recent past. The living giraffe (*Giraffa camelopardalis*) occurs in areas of high-quality foliage in the horseshoe-shaped area of grassland and woodland savannas in central Africa. Males may weigh close to 2000 kg and be over 5 m tall, enabling them to select high-quality browse from trees too high for other herbivores, which they gather using a tongue that may be almost 50 cm in length. The structures on the heads of living giraffes differ from the horns of antelopes and the antlers of deer in being formed initially as cartilage in the skin and then becoming ossified, but they remain covered with skin.

The characteristic patterning of the coat shows a great number of individual and regional variations, indicating subspecies (figure 3.80). Social structure may include herds of perhaps a dozen or so individuals, usually females and young, with mature males tending to be solitary or attaching themselves to a herd for no more than a few days at a time. Predators learn to avoid giraffes, which can move fast and deliver a lethal kick if provoked, although several populations of lions have developed methods for killing adult individuals by exploiting local circumstances (figure 3.35). Tarred roads in Kruger National Park in South Africa provide little traction for running giraffes, so hunting them there can be effective, while aardvark holes in Chobe National Park in Botswana have been seen to catch the legs of unwary animals and bring them down.

The okapi (*Okapia johnstoni*) is a much smaller and shorter animal than the giraffe, with a body weight up to 250 kg and a height just under 2 m, although it still has a rather elongated neck (figure 3.81). Today okapi are confined to areas of northeastern central Africa in the denser undergrowth of the rain forests, although not in the wet habitats. The okapi, like the giraffe, has a long and well-developed tongue, enabling it to browse selectively.

Miocene giraffids were a diverse group, comparatively strange looking to modern eyes, with large bodies but shorter necks than the extant species and often bizarrely shaped ossicones, as seen in the genus *Sivatherium* (figure 3.81). Initially browsers but secondarily adapted to grazing, although not specialists, species of *Sivatherium* coexisted with the earlier representatives of the modern-looking giraffes (figures 3.82 and 3.83). The long neck of the giraffe, the subject of considerable evolutionary debate over the years and seemingly the product of intensive selection for feeding on foliage high in trees, has been reconsidered by Robert Simmons and Lue Scheepers, who suggest selection for fighting ability in males as a more likely explanation.

Family Bovidae: Antelopes and Domestic Cattle, Sheep, and Goats

The antelopes, which make up the vast majority of the Bovidae, conjure up in people's minds the idea of a gentle, graceful, gazelle-like creature. But while many members of the family have just that appearance and perhaps nature, the African species range in stature from the hare-size dwarf antelopes of the tribe Neotragini to the massive buffalo and eland of the subfamily Bovinae. Neither are they all gentle; African buffaloes seek out and attempt to kill lion cubs hidden in bushes or longer grass by their mothers, and act aggressively in the face of threats from adult lions to the extent that they may recover calves from the cat's clutches.

The ecologist Norman Owen-Smith has pointed out that an exceptional feature of the living African bovid fauna lies in the diversity of grazing species, significantly greater than that on other continents, although once matched by the Miocene horses of North America. Among small to medium-size species, 17 are strict grazers, while a further eight are mixed grazers and browsers, and 12 are browsers. The origins of such diversity and dichotomy are most logically related to the extent of woodland and grassland available in Africa, with C_4 grasses providing nutrition along with a high-fiber content. Even among ruminants, grazers and browsers have different digestive capabilities and quite marked differences in the precise makeup and proportions of their digestive systems to enable them to cope with different foods.

The antelopes are the most diverse of the living ungulates, with more than 100 species in 44 genera, 72 of which are indigenous to Africa. Individual species of antelope are fairly habitat-specific, with their preference for the height and spacing of wood cover and its relationship to grassland quite well defined, as shown in studies undertaken by the paleontologist Elisabeth Vrba. Thus the distribution of suitable vegetation and conditions ensures the widespread distribution of some taxa and the

---

FIGURE 3.79

Sequential reconstruction of *Orangemeryx hendeyi*, drawn to scale with *Litocranius walleri*

The climacoceratid *Orangemeryx hendeyi* recently was described on the basis of abundant material from Arrisdrift in Namibia. It was a long-limbed, long-necked ruminant that probably browsed from sclerophyllous vegetation, as does the modern gerenuk (*Litocranius walleri*) (*bottom left*). But the most striking detail of its anatomy to a modern observer would be the strange, complex bony appendages on its head. Reconstructed shoulder height: 1.2 m.

FIGURE 3.80

Living giraffe subspecies

All nine living giraffe subspecies are virtually identical in morphology, and they are distinguished on the basis of their coat patterns. These photographs show near extremes of giraffe subspecies. On the coat of the reticulated giraffe (*Giraffa camelopardalis reticulata*) (*top*), the light background is reduced to thin lines between solid brown areas. This subspecies is typical of the semiarid Horn of Africa and is seen here in Samburu National Park, Kenya. The southern savanna giraffe (*G. camelopardalis giraffa*) (*bottom*), photographed in the Savuti Marsh of Chobe National Park, Botswana, displays a more typical coat pattern, with broken spots on a more extensive light background. Also evident is the sexual dimorphism typical of giraffes, with the smaller, lighter-colored female galloping in the foreground. (Photographs by Mauricio Antón)

FIGURE 3.81

Comparison of *Giraffa camelopardalis*, *Sivatherlum maurisium*, **and** *Okapia johnstoni*, **drawn to scale**

In spite of some differences in detail, the modern okapi (*Okapia johnstoni*) (*front*) has body proportions broadly similar to those of the ancestral members of the family Giraffidae, such as some species of the Miocene genus *Palaeotragus*. From such a generalized design, giraffids evolved into two contrasting body plans. On the one hand, the giraffines, like the modern giraffe (*Giraffa camelopardalis*) (*rear*) and several fossil relatives, developed very long necks and limbs, especially the forelimbs, which gives them a markedly sloping back. On the other hand, the sivatherines, like *Sivatherium maurisium* (*center*), evolved massive bodies more like those of large bovines, with robust, shortened limbs and relatively short necks. Within the genus *Sivatherium*, a trend is observed toward progressive shortening of the metacarpals (cannon bones of the forelimbs) relative to the metatarsals (cannon bones of the hind limbs). This trend may be associated with the increasingly grazing habit of derived sivatheres, in contrast to their mostly browsing ancestors. Apart from stout bodies, sivatheres are recognized by their spectacular cranial bony appendages, or ossicones.

Reconstructed shoulder height of *Sivatherium maurisium*: 1.72 m.

FIGURE 3.82

Comparison of the heads of *Giraffa jumae* and *Giraffa camelopardalis*

Known from good fossil remains from Rawe, Olduvai Gorge in Tanzania, and East Turkana in Kenya, among other sites, *Giraffa jumae* (*bottom*) had attained the size and characteristic body proportions of the modern giraffe (*G. camelopardalis*) (*top*), but its skull retained primitive features reminiscent of earlier paleotragine giraffids. Compared with the skull of a modern giraffe, that of *G. jumae* was lower, and it almost completely lacked the protuberance in front of the orbits, while the supraorbital appendages, well developed as in the modern species, were more posteriorly inclined.

FIGURE 3.83

Life reconstruction of *Giraffe jumae*

The bulk of the diet of modern giraffes consists of leaves of trees of the families Combretaceae and Mimosaceae, and the same is likely to have been true for extinct species. But giraffes eat from a great variety of plants (more than 100 species recorded) and are among the few animals to go for the hard, toxic fruit of the sausage tree (*Kigelia africana*). This tree is recorded as a fossil as early as the Pliocene in the Omo River of Ethiopia, so it is likely that the extinct *Giraffa jumae* would have browsed for its fruit, just like its extant relatives. The coat pattern in this reconstruction follows a hypothesis that the pattern of modern giraffes evolved from a primitive dark background with light spots that progressively grew in star-like shape, until the light areas eventually joined and the dark areas became isolated as islands or spots. This scene shows *G. jumae* at an early stage of that hypothetical process, with dark areas still only partially surrounded by light ones.

Reconstructed total height: 5 m.

restricted distribution of others. In addition, closely related species—that is, members of the same tribe—show a broadly similar range of habitat preferences. Since the family is too large to discuss individual species in detail here, we instead offer a brief summary of the subfamilies and tribes and mention some of their members. It is worth stressing that the rank of tribe is particularly useful as a means of grouping the numerous species of African antelope when dealing with fossil material. Doing so allows us to correlate the appearance of certain types of antelope in the fossil record with larger-scale changes in climatic conditions and vegetational structure.

Subfamily Hippotraginae

The Hippotraginae is a large subfamily, with around 20 living species in four tribes.

Tribe Alcelaphini: Wildebeests and Hartebeests

The Alcelaphini are medium-size to large antelopes with long faces and legs and often strangely shaped horns (figure 3.84). Typical species in the tribe are the blue (*Connochaetes taurinus*) and black (*C. gnou*) wildebeests, at 300 and 200 kg, respectively, the heaviest; topi (*Damaliscus lunatus*); blesbok (*D. dorcas*); kongoni or hartebeest (*Alcelaphus buselaphus*); and hirola (*Beatragus hunteri*), which now is confined to a small range in Kenya. The alcelaphines are grazers with a general preference for open vegetation and tend to have wide muzzles that allow them to eat short and nutritious grasses. Figure 3.85 shows some living and fossil alcelaphines.

Tribe Hippotragini: Oryxes

The Hippotragini—literally, horse-like antelopes—are also large grazers with long horns, although better adapted than the Alcelaphini to the arid conditions of northern, eastern. and southwestern Africa, exemplified by the fact that their range has included the Arabian Peninsula (figure 3.86). The tribe includes the roan (*Hippotragus equinus*) and sable (*H. niger*) antelopes, which have the narrow muzzles of specialist feeders; the scimitar-horned (*Oryx dammah*), beisa (*O. beisa*), and southern (*O. gazella*) oryxes; and the addax (*Addax nasomaculatus*). Both alcelaphines and hippotragines are well adapted to run and thus to flee from predators.

Tribe Reduncini: Bucks

The Reduncini consists of three species of reedbuck of the genus *Redunca*, weighting 40 to 95 kg, and five species of *Kobus*, medium-size to large antelopes, all of which occur in better-watered areas (figure 3.87). The biggest of the *Kobus* species, and the one with the widest distribution, is the waterbuck (*Kobus ellipsiprymnus*), at 300 kg; the smallest, the kob (*K. kob*), at 77 kg, is found in a band from Senegal to Lake

Victoria. The lechwe (*K. leche*) has particularly splayed hooves that enable it to live in waterlogged areas of central Africa, to which it is confined by its need to drink. Reduncines generally seek more specialist grasses and have narrow muzzles, a feature evident in *Menelikia lyrocera* (figure 3.88).

Tribe Peleini: Rhebok

The Peleini consists of a single South African species, the rhebok (*Pelea capreolus*), a small, long-legged antelope that mainly inhabits upland areas where it grazes.

The Hippotraginae, particularly the Alcelaphini and Hippotragini, have a considerably more diverse fossil record than living species numbers would imply. The alcelaphines first clearly appear at around 5 Ma in the form of the more primitive species *Damalacra acalla* and *D. neanica* at Langebaanweg in the Cape Province of South Africa. However, this datum may be pushed back slightly when material from Lothagam in Kenya is described in more detail, identifications are published and confirmed, and dates are more clearly established. The hippotragines appear a little earlier, at around 6.5 Ma, with unnamed species at Lothagam, and the Mpesida Beds of the Baringo Basin have produced the reduncine *Dorcadoxa prorrecticornis*. A major radiation occurred between 3 and 2 Ma, centered on 2.5 Ma. At that time, many of the living species first appear, together with members of the Antilopini, as well as such larger animals as the extinct giant hartebeests and hirolas of the genera *Megalotragus* and *Beatragus* and larger hippotragines like *Hippotragus gigas*. This radiation also includes the first evidence of the lineage of the largest reduncine, the waterbuck.

Subfamily Aepycerotinae

Tribe Aepycerotini: Impalas

There is only one living species of impala (*Aepyceros melampus*), a medium-size antelope that weighs up to 80 kg and has a wide distribution in eastern and southern Africa, where it is generally associated with grassland to woodland ecotones (figure 3.89). The impala has a generally gazelle-like appearance, although all authorities consider it to be completely separated from the true gazelles, and it has been placed as a sister group to the Alcelaphini. It is an adaptable, intermediate feeder, able to subsist in a variety of habitats, and in some game parks it may make up more than 70 percent of the bovids.

The fossil record of impalas shows an interesting pattern of little or no diversification over time since the first appearance of the Aepycerotini at around 6.5 Ma, in comparison with the extensive radiation of its sister group (figure 3.90). Elisabeth Vrba has suggested that the clue to this difference lies in the more generalized nature

FIGURE 3.85

Comparison of *Alcelaphus buselaphus*, *Parmularius altidens*, and *Megalotragus issaci*

The modern hartebeest (*Alcelaphus buselaphus*) (*left*) exemplifies the typical features of the tribe Alcelaphini at their most evolved: it is a large antelope, with long legs, high shoulders, and a sloping back; a rather short and muscular neck; a very long face; and twisted horns. These features are regarded as adaptations for a gregarious life as a grazer in open grassland: the limb proportions are ideal for sustained fast locomotion, the short neck provides strength for frequent horn fighting, and the long head compensates for the shortness of the neck by allowing the hartebeest to reach short grasses. *Parmularius altidens* (*center*) was a slightly smaller antelope, and alcelaphine features were still moderate, somewhat resembling those of the modern topi. Abundant postcranial remains from Olduvai Gorge in Tanzania show that the forelimbs of *Parmularius* were not as elongated as those of the hartebeest and topi, and thus the animal did not have as pronounced a sloping back. *Megalotragus* (*right*), though, was more specialized than the extant hartebeest in some features, especially its larger body size (up to 1.4 m at the shoulder) and the remodeling of the skull.

Each square measures 50 cm on a side.

FIGURE 3.84

Living alcelaphine antelopes

The topi (*Damaliscus lunatus*) (*top*), such as this one in the Masai Mara National Reserve, Kenya, is a typical alcelaphine antelope with a sloping back, a long head, and horns that have backward-curving stems and forward- and inward-curving tips. The hartebeest (*Alcelaphus buselaphus*) (*center*), also in Masai Mara, takes the alcelaphine adaptations one step further, with a very long head and strongly lyrate horns stemming from a pedicle. The wildebeest (*Connochaetes taurinus*) (*bottom*), like this one in Etosha National Park, Namibia, differs from the other alcelaphines in its shorter legs, thicker neck with mane and beard, and vaguely cow-like horns. (Photographs by Mauricio Antón)

of the impala's feeding niche, which exists independently of environmental changes. If impalas can cope in a range of circumstances, they are less likely to suffer range fragmentation over time and the kinds of genetic change, including speciation, that may result. The rather more specialist alcelaphines and hippotragines, in contrast, have seemingly encountered such habitat fragmentation, and their radiation is the effect of the difference in specialization between them and the impala.

Subfamily Antilopinae
Tribe Neotragini: Dwarf Antelopes
The Neotragini—dwarf antelopes like the klippspringer (*Oreotragus oreotragus*), the dik-diks (*Madoqua kirki* and various others) (figure 3.91), and such members of the genus *Raphicerus* as the steenbok (*Raphicerus campestris*) and grysbok (*R. melanotis*)—weigh between under 1 and perhaps 22 kg and are mainly browsers in more closed habitats. The most widely distributed is the klippspringer, one of the largest at up to 18 kg. The oribi (*Ourebia ourebi*), the largest member of the tribe and a grazer, is also widely distributed across the central horseshoe of grassland savanna.

Tribe Antilopini: Gazelles
The Antilopini, or gazelline antelopes, consists of three or perhaps four genera, of which the most numerous is *Gazella*, with eight species (figures 3.92 and 3.93). They are small (15–25 kg) to medium-size (30–50 kg) animals with long legs and necks, generally adapted to coping with heat and aridity, and are found across northern and northeastern Africa and, of course, across Arabia and into Asia. The dibatag (*Ammodorcus clarkei*), a somewhat larger animal, has a highly restricted distribution in Somalia and Ethiopia, where it feeds on what is known as camel brush. The gerenuk (*Litocranius walleri*) inhabits the semiarid bushland over much of northeastern Africa. It is one of the largest and tallest of the gazelles, with a small and pointed muzzle, and by rising on its hind legs is able to take selective elements of browse at a height of around 2 m. The last species, the springbok (*Antidorcas marsupialis*), is found only in southwestern Africa, where it behaves in a generally gazelle-like manner. However, some authorities differentiate it from the true Antilopini on

FIGURE 3.86
Living hippotragine antelopes

(*Top*) Sable antelope (*Hippotragus niger*), Victoria Falls National Park, Zimbabwe. (*Center*) Beisa oryx (*Oryx beisa*), Samburu National Park, Kenya. (*Bottom*) Gemsbok or southern oryx (*O. gazella*), Etosha National Park, Namibia. Not all experts agree that the oryx and gemsbok belong to separate species, but the differences in build and markings (with the gemsbok being heavier and having thicker black markings) are consistent and at least indicative of the long isolation between the two populations. (Photographs by Mauricio Antón)

the basis of differences in dental structure and horn-core architecture, placing it in a separate tribe: the Antidorcini. It is an adaptable animal, able to exist both by grazing and by browsing and to obtain sufficient moisture from its food alone.

Fossil antilopines, including springboks, are quite well known, with the tribe represented from the mid Miocene onward, while the reduncines can be traced back for only around half that time to the latest Miocene. Fossil species of the genus *Gazella* are known from deposits at Fort Ternan in Kenya dated to 15 Ma, and the distribution

FIGURE 3.88

Life reconstruction of *Menelikia lyrocera*

This reconstruction of the reduncine antelope *Menelikia lyrocera* is based on cranial and postcranial material from the Shungura Formation at the Omo River, Ethiopia.
Reconstructed shoulder height: 1 m.

FIGURE 3.87

Living reduncine antelopes

(*Top*) Puku (*Kobus vardoni*), Chobe National Park, Botswana. (*Center*) Lechwe (*K. leche*), Okavango Delta, Botswana. (*Bottom*) Defassa waterbuck (*K. ellipsiprymnus defassa*), Lake Nakuru National Park, Kenya. (Photographs by Mauricio Antón)

FIGURE 3.89
Impala

This handsome male impala (*Aepyceros melampus*), at Pilanesberg National Park, South Africa, is surrounded by the females and young of his harem. (Photograph by Mauricio Antón)

of the extinct *Antidorcas recki* (figure 3.94) included both eastern and southern regions of Africa during the later Pliocene.

SUBFAMILY BOVINAE
Tribe Bovini: Buffalo
The only living member of the Bovini in Africa is the buffalo (*Syncerus caffer*), a large browsing and grazing antelope (figure 3.95). Males, which may weigh up to 1000 kg, have massively developed horns that spread sideways from a large cranial boss. A smaller forest form is also known. The buffalo has a wide distribution outside the most arid areas.

Tribe Tragelaphini: Nyalas, Kudus, Bushbucks, and Elands
The Tragelaphini, the spiral-horned antelopes, comprise seven species of nyala, kudu, and bushbuck and two species of eland. They range in size from the bushbuck (*Tragelaphus scriptus*), at up to 80 kg; through the greater kudu (*T. strepsiceros)* and bongo

FIGURE 3.90

Life reconstruction of the head of *Aepyceros premelampus*

This reconstruction of the head of *Aepyceros premelampus* is based on the holotype, a partial skull with horn cores found at Lothagam in Kenya, and shows the great resemblance between this early impala species and its modern counterpart (*A. melampus*). The main differences are the smaller size and more gracile appearance of the fossil species, and the less pronounced curvatures of its horns.

(*T. euryceros*), at 300 to 400 kg; to the Cape (*Taurotragus oryx*) and Derby's (*T. derbianus*) elands, which rival the buffalo in size (figure 3.96). All eat browse, although elands may take up to 40 percent of their diet from grasses, and the tribe is widely distributed in sub-Saharan Africa, although individual species are more localized. Thus the long-legged sitatunga (*Tragelaphus spekei*) and the short-legged bongo are generally restricted to the drainage basin areas of western central Africa. The nyala (*T. angasi*) and the lesser kudu (*T. imberis*) are restricted to the southeastern and northeastern parts of the continent, respectively, while the mountain nyala (*T. buxtoni*) has the most restricted range of all: the southeastern highlands of Ethiopia.

Fossil Bovinae in Africa were much more abundant than living bovids, especially among the Bovini after their first appearance at around 6 Ma. From around 2.5 Ma, they included members of the large and bizarrely adorned genus *Pelorovis*, whose horn cores show an animal with horns far in excess of anything seen in the living African buffalo. *Pelorovis oldowayensis* (figure 3.97) and *P. turkanensis* have wide,

FIGURE 3.91
Damara dik-dik

Male and female Damara dik-diks (*Madoqua kirkii*), Etosha National Park, Namibia. (Photograph by Mauricio Antón)

downward- and backward-sweeping horn cores, while those *of P. antiquus* (figure 3.98) are more reminiscent of the horns of the modern Asian water buffalo. Tragelaphini are known from slightly earlier, at around 6.5 Ma (figure 3.99). The African fossil record also contains Miocene representatives of the tribe Boselaphini, now found in Asia. Boselaphines like *Protragoceros* were likely the earliest of all bovids to enter Africa (figure 3.100). Perhaps the last African record is at Langebaanweg in the Cape Province of South Africa in the form of *Mesembriportax acrae*.

SUBFAMILY CEPHALOPHINAE
Tribe Cephalophini: Duikers
The duikers are essentially small (below 25 kg) forest-dwelling antelopes, placed in two genera. *Sylvicapra* contains the single species *Sylvicapra grimmia*, the bush duiker,

FIGURE 3.92
Living antilopine antelopes

(*Top*) Male and female Grant's gazelle (*Gazella granti*), Samburu National Park, Kenya. (*Center*) Male Thomson's gazelle (*G. thomsoni*), Masai Mara National Reserve, Kenya. (*Bottom*) Male springbok (*Antidorcas marsupialis*), Etosha National Park, Namibia. (Photographs by Mauricio Antón)

FIGURE 3.93

Two aspects of behavior of the gerenuk

(*Top*) A male gerenuk (*Litocranius walleri*) scratches an ear while guarding his harem and young.
(*Bottom*) A female stands on her hind legs while browsing for high branches. (Photographs by Mauricio
Antón)

FIGURE 3.94

Sequential reconstruction of the head of *Antidorcas recki*

This reconstruction of the head of the antilopine antelope *Antidorcas recki*, based on well-preserved cranial fossils from Olduvai Gorge in Tanzania, shows the overall similarity to the head of the modern springbok (*A. marsupialis*), although the shape of the horns was different, with a clear backward curvature.

which is widely distributed in savannas and lighter woodlands (figure 3.101). *Cephalophus* contains 16 species of forest duiker, which among them live in most African woodlands. Only Jentink's duiker (*Cephalophus jentiki*), an inhabitant of Sierra Leone forests, reaches any great size, with a weight range up to 80 kg.

The Cephalophini have a patchy African fossil record as far back as the Miocene, but appear to have been always confined to that continent.

SUBFAMILY CAPRINAE

Tribes Ovibovini and Caprini: Goats

The Caprinae consists of two African species: the Barbary sheep (*Ammotragus lervia*) and the Nubian ibex (*Capra ibex*). The sheep, in which the highly decorative males may weigh up to 140 kg, were distributed over much of northern Africa in historic times, but now occur in isolated populations in the highland areas of the Sahara. The slightly smaller ibex, essentially a Eurasian species, is restricted to highland areas to the east of the Nile in Sudan and to the Simen Mountains of Ethiopia, where its superb adaptations to climbing seemingly featureless rock allow it to evade most predators.

The African fossil record of the Caprinae is much more extensive than the number of living species would suggest. The subfamily is Eurasian in origin, and its members

FIGURE 3.95

Buffalo

Male and female buffalo (*Syncerus caffer*), Samburu National Park, Kenya. (Photograph by Mauricio Antón)

FIGURE 3.96

Living tragelaphine antelopes

(*Top*) Bushbuck ram (*Tragelaphus scriptus*),Victoria Falls National Park, Zimbabwe. (*Center*) Cape eland (*Taurotragus oryx*), Etosha National Park, Namibia. (*Bottom*) Kudu bull (*Tragelaphus strepsiceros*), Etosha National Park, Namibia. (Photographs by Mauricio Antón)

appear to have dispersed in Africa at various times from the mid Miocene onward. There was a particularly marked incursion at around 2.5 Ma, when members of both the Ovibovini and the Caprini appear in eastern and southern regions of the continent, with the ovibovine *Makapania broomi* at Sterkfontein and Makapansgat Limeworks in South Africa and various unnamed caprines in deposits of the Shungura Formation and West Turkana in eastern Africa. Yet more unnamed members of both tribes also appear in the latter deposits just after 2 Ma.

In general, the Bovidae is extensively represented in the African fossil record from Miocene times around 18 Ma onward, when the antelopes appear to have immigrated to the continent. Some 147 individual taxa have been identified in the fossil and extant records over the past 7 Ma alone; not surprisingly, their remains may account for up

Life reconstruction of *Tragelaphus nakuae*

This reconstruction of *Tragelaphus nakuae*, based on skull fragments, horn cores, and postcranial remains from the Omo River in Ethiopia and the Koobi Fora Formation of East Turkana, Kenya, shows that this tragelaphine antelope would have resembled some of the more robust extant species of the genus *Tragelaphus*, such as the bongo (*T. euryceros*).

Reconstructed shoulder height: 1.22 m.

Life reconstruction of *Pelorovis oldowayensis* in two views, compared with *Syncerus caffer*

The best fossils of *Pelorovis oldowayensis* were found, as its name suggests, at Olduvai Gorge in Tanzania. There, not only an associated skeleton but also the remains of what appears to have been a herd were recovered. Compared with the modern buffalo (*Syncerus caffer*) (*right*), *P. oldowayensis* was of broadly similar size, but the individual limb bones, particularly the distal segments, were considerably longer, making the bovine taller. But the greatest differences were in the skull. It is very elongated in a way reminiscent of the skulls of alcelaphine antelopes, a feature that suggests a similar habit of grazing short grass. The horns (enormous in the males)—which project backward, then outward, and finally forward—are unlike those of any other bovine and led early paleontologists to classify this animal as a member of the tribe Ovibovini.

Reconstructed shoulder height: 1.55 m.

Life reconstruction of *Pelorovis antiquus*

A complete, very well preserved skeleton of *Pelorovis antiquus*, found near Djelfa in Algeria in the late nineteenth century, remains the best guide for the reconstruction of the life appearance of this bovine. Like *P. oldowayensis*, it was about the size of a modern buffalo, but instead of having elongated distal limbs, *P. antiquus* was as robust as or even more so than the living animal. The skull was not nearly as elongated as that of *P. oldowayensis*, and the horns, while proportionally enormous, had a curvature more like that of the horns of the familiar extant Asian water buffalo, a similarity that led early specialists to classify the fossil species as a member of the same genus, *Bubalus*. In view of the overall appearance of the reconstructed animal, one can understand such a choice, because its similarity to the water buffalo is much greater than its resemblance to the other *Pelorovis* species.

Reconstructed shoulder height: 1.64 m.

FIGURE 3.100

Life reconstruction of *Protragoceros labidotus*

Well-preserved skulls and postcranial fossils from the site of Fort Ternan in Kenya show that *Protragoceros labidotus* had the proportions of a woodland dweller, with the hind limbs relatively longer than the fore-limbs. This drawing shows a male of the species; females were hornless.
Reconstructed shoulder height: 67 cm.

to 80 percent of the specimens of larger animals in fossil localities. Because of their present-day abundance, the bovids often form an equally large proportion of the prey of larger predators, and the survival of such a range of large predators in Africa probably is related to the presence of the antelopes as a food source on that continent.

Order Insectivora: Insect Eaters

The diverse and species-rich Insectivora contains the families of common shrews (Soricidae), otter shrews and tenrecs (Tenrecidae), hedgehogs (Erinaceidae), and golden moles (Chrysochloridae). Among these insectivores, the tenrecs and golden moles are long-established members of the endemic African fauna. The distribution patterns of members of this order vary; some of the moles and shrews are quite restricted, with the tenrecs confined to Madagascar, while the African hedgehogs, with four species in the genus *Atelerix* and two in the genus *Hemiechinus*, occur widely in

FIGURE 3.101
Bush duiker

A bush duiker (*Sylvicapra grimmia*) peeks out from behind grasses in Kruger National Park, South Africa. (Photograph by Mauricio Antón)

more arid habitats in the savannas and along the Mediterranean littoral. White-toothed shrews, with more than 100 species in the genus *Crocidura*, inhabit most of Africa.

The tenrecs underline the problems thrown up by the biomolecular approaches to determining relationships between organisms and thus their assignment to taxonomic groups. The tenrecs and the golden moles are placed by these analyses in the Afrotheria, while the otter shrews, now confined to western central Africa, are not. Yet conventional, morphologically based studies of relationships put the otter shrews and the tenrecs in the same family: the Tenrecidae.

Whatever the taxonomic problems, the fossil record of the insectivores, as those of the rodents and all small mammals, is extremely patchy and heavily dependent on accumulations of food debris from predator feeding. The life appearance of a fossil hedgehog (*Gymnurechinus camptolophus*) has been reconstructed, though, based on remains from Rusinga Island in Lake Victoria, Kenya (figure 3.102).

FIGURE 3.102

Life reconstruction of *Gymnurechinus camptolophus*

Several details of the anatomy of this early hedgehog from Rusinga Island in Lake Victoria, Kenya, show that it would have resembled the modern hairy hedgehogs, such as *Echinosorex*, rather than the typical, spiny hedgehogs like *Erinaceus* and *Paraechinus*. Among other features, *Gymnurechinus camptolophus* shows a long neck and strong muscular attachments in the vertebral column. Spiny hedgehogs, in contrast, have specialized skin musculature that helps roll them into a spiny ball, and thus takes over some of the flexing functions of the spinal musculature, which, in turn, is reduced.
Reconstructed head and body length: 25 cm.

FIGURE 3.103

Life reconstruction of *Rhynchocyon rusingae*

The fossil macroscelidian *Rhynchocyon rusingae* would have been very similar to the living giant elephant shrews from the forests of central Africa. A coat pattern with a series of rows of light-colored spots on a brownish background appears to be the primitive condition for members of this genus, so we have attributed such a coat to the fossil species from Rusinga Island in Lake Victoria, Kenya.

Order Macroscelidea: Elephant Shrews

The Macroscelidea contains a single family, the Macroscelididae, of a dozen or so species of African shrew that originally were classified as a family within the Insectivora, but now are thought by some to be closer in evolutionary relationship to the rodents. Again, the elephant shrews appear to be members of the Afrotheria.

The earliest known representatives of the order are found in the Oligocene deposits of the Fayum Depression in Egypt, but specimens are also known from early and mid Miocene deposits in Namibia and Kenya (figure 3.103). Members of the fossil subfamily Myohyracinae formerly were regarded as hyracoid, as the name suggests. *Myohyrax oswaldi*, known from almost complete cranial and mandibular remains as well as from postcranial bones from the early Miocene of Kenya, is the most completely known fossil species of the order.

Order Chiroptera: Bats

Bats make up one of the most diverse mammalian orders, which, with around 1000 species worldwide, includes almost one-quarter of the living mammalian fauna (figure 3.104). The Chiroptera is conventionally subdivided into the suborder Megachiroptera, with a single family of fruit-eating bats, and the suborder Microchiroptera, with several families of insect-eating species. The living African Megachiroptera comprise 25 to 30 variously sized species, some quite large, while the very diverse Microchiroptera, with more than 170 species, are generally small. Many insectivorous bats are widely distributed and occur in vast numbers, although fruit bats are more restricted in habitat requirements and therefore less widely found.

Bats are something of a mammalian oddity. Their bodies are well adapted for flight, and, as a result, many of their bones are thin, light, and extremely fragile. Thus the fossil record is extremely patchy; there are beautifully preserved specimens of Eocene age from shale deposits at Messel in Germany, remains of animals that fell into a lake and were preserved in the lake sediments. Such finds give a clear and unequivocal picture of the early evolution of the order, although elsewhere relatively little material is available. What we see from the Eocene material is that bats achieved their highly specialized level of development at an early stage and that they have undergone relatively little structural evolution since then.

FIGURE 3.104

Life reconstruction of *Hipposideros*

The known fossil remains of species of *Hipposideros*, such as those from Songhor in Kenya, are very fragmentary, but they are enough to show that this bat would have resembled closely its extant relatives, the members of the family Hipposideridae. These bats have relatively short, wide wings adapted for flight among dense vegetation as well as characteristic, leaf-like appendages in their noses that serve to focus and concentrate the sound waves emitted through their nostrils.

Order Pholidota: Pangolins

Pangolins, or scaly anteaters, placed in the single family Manidae in the Pholidota, are among the strangest of the mammals, with their brown, armor-like scales and their habit of rolling into a defensive ball (figure 3.105). Perhaps the best comparison is with the armadillos of the New World, although they belong to a different order, the Edentata, but the biomolecular analyses that identify the Afrotheria have suggested a closer affinity with the Carnivora. Four living species occur, ranging in size from that of a marten to that of a badger, with the largest weighing up to 35 kg. The long-tailed (*Uromanis tetradactyla*), white-bellied (*Phataginus tricuspis*), and giant ground (*Smutsia gigantea*) pangolins have similar distributions in western African rain forests, while the Cape, or Temminck's ground, pangolin (*S. temminckii*) is found in dry bushland of eastern and southern regions. All live on ants and termites.

Fossil pangolins are known in Africa from Pliocene deposits in Uganda, late Miocene deposits at Langebaanweg in the Cape Province of South Africa, and the mid Miocene locality of Serek in Kenya.

FIGURE 3.105
Cape pangolin
The large, widespread Cape, or Temminck's ground, pangolin (*Smutsia temminckii*) walks on two legs when foraging, occasionally touching the ground with its tail or forelegs to counterbalance.

Order Lagomorpha: Hares and Rabbits

African representatives of the Lagomorpha, all relatively recent immigrants, fall into three genera—*Lepus* (true hares), *Pronolagus* (rock hares), and *Bunolagus* (riverine rabbit)—all placed in one family: the Leporidae (figure 3.106). Most are quite widely distributed in more open areas. The common rabbit (*Oryctolagus cuniculus*) is found in northern Morocco and Algeria, but it is quite likely that it was introduced by human activity even more recently.

Order Rodentia: Rodents

The Rodentia, members of which are characterized by their well-developed incisor teeth, contains the families of mole rats (Bathyergidae), porcupines (Hystricidae) (figure 3.107), springhares (Pedetidae), dormice (Gliridae), squirrels (Sciuridae), and gliding rodents, or scaly-tailed flying squirrels (Anomaluridae), as well as the super-family of gerbils, rats, and mice (Muroidea), all with differing distributions in Africa. The order is one of the most successful in evolutionary terms, with more than 1600

FIGURE 3.106

Riverine rabbit

The common name is slightly misleading, since the riverine rabbit (*Bunolagus monticularis*) is considered to be a hare rather than a rabbit. Ironically, while the European common rabbit has become a plague in several parts of the world, the riverine rabbit is one of the most endangered mammals of Africa.

FIGURE 3.107

Cape porcupine

Porcupines, such as the Cape, or South African, porcupine (*Hystrix africaeaustralis*), are very large rodents that weigh up to 24 kg. They inhabit a wide spectrum of habitats, including relatively open areas where they roam relatively safe from predators thanks to their heavy armor of spines. Porcupines have the habit of gnawing bones, leaving characteristic markings, so their presence often can be documented at fossil sites even if their own bones are not in evidence.

species, or about one-third of living mammals. But the fossil record, as for many smaller mammals, is extremely patchy, although it does throw up some oddities. An interesting early Miocene form is the springhare (*Parapedetes namaquensis*), known from Elisabethfeld in Namibia (figure 3.108). From Lothagam in Kenya, we even have a giant squirrel (*Kubwaxerus pattersoni*) (figure 3.109). Localized abundance may occur in some cave deposits and undoubtedly is the result of pellet regurgitation by birds of prey such as owls. These fossil assemblages can be extremely rich, and if the sediments were later cemented by groundwater minerals, the result can look like a solid rock composed of rodent remains.

Many rodent taxa are habitat-specific, and the presence of their remains in a deposit may offer numerous clues to past environmental conditions. Assemblages produced by owl pellets may provide a rich variety of species, but care must be exercised in interpreting such fossils since the predatory behavior and selection of the owl may be a major factor in the structuring of the samples.

Order Tubulidentata: Aardvarks

The aardvark (*Orycteropus afer*) is the sole living representative of the Tubulidentata and is widely distributed in sub-Saharan Africa. Members of the order are reported from Miocene localities in the eastern part of the continent. Although Miocene occurrences also are known in Eurasia, particularly in Greece, an African origin is strongly implied and now is strengthened by the inclusion of aardvarks in the Afrotheria. The genus *Orycteropus* is known from the late Miocene, where it is represented by the smaller and less robust *O. gaudryi*, although material from early Miocene deposits has been referred to the genus *Myorycteropus*, shown in comparison with the living aardvark in figure 3.110. From Lothagam in Kenya, the species *Leptorycteropus guilielmi* recently has been described (figure 3.111).

The Afrikaans name aardvark means "earth pig" in English and is entirely descriptive of the appearance and habits of the animal, which burrows in the ground and subsists variously on termites, termite fungi, and various other insects. It has a heavily built skeleton with large muscle attachments and is well equipped for digging. The teeth, which lack enamel, grow continuously.

Order Hyracoidea: Hyraxes

The Hyracoidea is represented by three rabbit-size species of hyrax, or dassie. The rock hyrax, or Cape dassie (*Procavia capensis*) (figure 3.112), inhabits much of sub-

FIGURE 3.108

Size comparison of *Parapedetes namaquensis*, *Pedetes capensis*, and *Megapedetes pentadactylus*

The early springhare (*Parapedetes namaquensis*) (*left*), from the lower Miocene of Elisabethfeld in Namibia, was smaller than the extant species (*Pedetes capensis*) (*center*) and shows a mixture of primitive and derived features that make it an unlikely direct ancestor of the living springhare. *Megapedetes pentadactylus* (*right*), from the lower Miocene of Songhor in Kenya, is the largest known member of the family Pedetidae, but was otherwise primitive in all respects relative to the extant species. Both fossil species display similar body proportions and other anatomical features, indicating an adaptation for hopping like that of the modern springhare. This form of locomotion combines speed with a great economy of energy. Reconstructed head and body length of *Megapedetes pentadactylus*: 50 cm.

FIGURE 3.109

Life reconstruction of *Kubwaxerus pattersoni*

The relatively complete sample of cranial and postcranial bones of *Kubwaxerus pattersoni* from Lothagam in Kenya allows a confident restoration of its body proportions, which are similar to those of the extant, closely related giant squirrels of the tribe Protoxerini, although it was considerably larger than any of the modern species. Several features of the postcranial skeleton indicate that this squirrel, although largely arboreal, would have spent a significant amount of its time foraging on the ground, as shown in this drawing.

Reconstructed head and body length: 37 cm.

FIGURE 3.110

Comparison of *Myorycteropus africanus* and *Orycteropus afer*

This reconstruction of *Myorycteropus africanus* (*front*) is based on a partial skeleton and other isolated fossils from Rusinga Island in Lake Victoria, Kenya. As the illustration shows, *Myorycteropus* was much smaller than the extant aardvark (*Orycteropus afer*), but recognizably aardvark-like nonetheless. The bones of the forelimb are proportionally even more robust than those of the modern species, indicating a well-developed digging ability, while the hind limbs are proportionately more gracile. In spite of its small size, *Myorycteropus* was no doubt able to break through termite mounds in much the same way as its modern relative does.

Saharan Africa except the rain forests of the Congo Basin, Zambia, and much of Botswana. In rain forests generally, it is replaced by the tree dassie (*Dendrohyrax arboreus*). Bruce's dassie (*Heterohyrax brucei*) lives in much of eastern Africa as well as parts of Angola and Botswana.

Hyraxes are placed in the Afrotheria. Ancestral forms were much larger than living species and different in appearance. They could be considered the dominant medium-size herbivore on the continent during the latest Eocene and into the Oligocene, making up one-half of the fauna in parts of the sequence in the Fayum Depression of Egypt. *Gigantohyrax maguirei*, from late Pliocene deposits at Makapansgat Limeworks in the Transvaal of South Africa, was three time the size of living forms, while the equally large *Prohyrax hendeyi*, from Arrisdrift in Namibia, had a markedly more upright stance as judged from the shape of its limb joints (figures 3.112 and 3.113).

FIGURE 3.111

Life reconstruction of *Leptorycteropus guilielmi*

The fossil aardvark *Leptorycteropus guilielmi* was described on the basis of a partial skeleton found at Lothagam in Kenya that shows *Leptorycteropus* as a much smaller and more lightly built animal than the modern aardvark (*Orycteropus afer*). The back was arched, as in the living species, but the limbs were longer and less muscular. The muzzle, as indicated by the length of the tooth row, would have been relatively shorter.

Reconstructed shoulder height: 24 cm.

FIGURE 3.112

Size comparison of *Prohyrax hendeyi* and *Procavia capensis*

Cranial and postcranial remains of the giant hyrax *Prohyrax hendeyi* (*rear*) are very abundant at the early to mid Miocene site of Arrisdrift in Namibia. It was much larger than the extant hyraxes, such as the rock hyrax (*Procavia capensis*), but the differences went beyond mere size. The long bones of the limbs have straighter shafts, and the articulations of the elbow, wrist, knee, and ankle show that the fossil hyrax moved with more upright limbs, in contrast with the crouching stance of the modern species, implying more efficient locomotion on dry ground.

Reconstructed shoulder height of *Prohyrax hendeyi*: 37 cm.

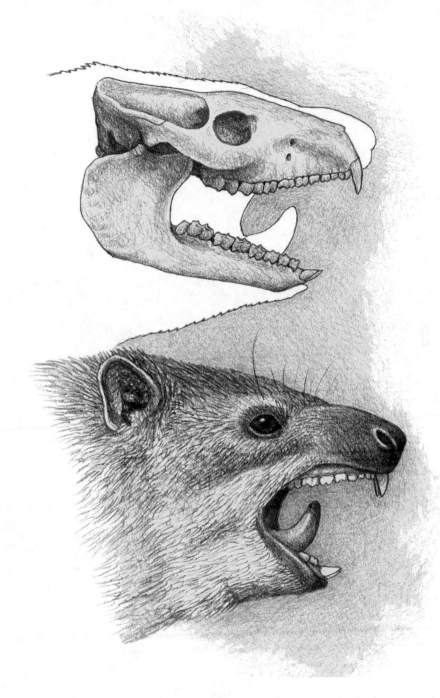

FIGURE 3.113

Life reconstruction of the head of *Prohyrax hendeyi*

This reconstruction of the head of *Prohyrax hendeyi*, based again on material from Arrisdrift in Namibia, shows the overall hyrax-like appearance of the skull and head of this giant hyrax. The main differences with the modern rock hyrax are the larger size, the orbits closed posteriorly, and the very short diastema behind the anteriormost incisor teeth.

PLATE 1

Woodland savanna with grassy patches in Chobe National Park, Botswana

Grassy woodland provides a good habitat for a host of browsers, such as these giraffes, and probably has been widespread in Africa since the Miocene. (Photograph by Mauricio Antón)

PLATE 2

Grassland savanna with shrubs and trees in Chobe National Park, Botswana

Grassland savanna encompasses a range of plant types that allow the coexistence of browsers, grazers, and such mixed feeders as these impala. (Photograph by Mauricio Antón)

PLATE 3

Woodland savanna with grassy patches on hilly ground in Pilanesberg National Park, South Africa

In this type of highveld savanna were the caves that have become the Pliocene and Pleistocene fossil sites in the Sterkfontein Valley. (Photograph by Mauricio Antón)

PLATE 4

Floodplain grassland and gallery woodland in the Okavango Delta, Botwsana

Many of the grasslands in Africa are maintained by the annual flooding of rivers, which prevents the growth of trees in low ground. As this aerial view of the delta shows, the higher ground is marked by the presence of riverine trees. (Photograph by Mauricio Antón)

Riverine woodland in the Okavango Delta, Botswana

Groves of palms, figs, and other large trees grow around the borders of the floodplains, as they grew along the fluvial systems that created many of the fossil sites, especially in eastern Africa. Fossils of tree species typical of such riverine woodland have been found in deposits of the Shungura Formation at the Omo River, Ethiopia, among others. (Photograph by Mauricio Antón)

PLATE 6

Grassland savanna in the Masai Mara National Reserve, Kenya

Extensive grassland provides excellent habitat for grazing ungulates, such as these common zebras, but it is largely artificial. Rainfall in the Masai Mara region is high enough to maintain a woodland savanna, more like those in plates 1 and 2, but the area is burned annually by Masai pastoralists to create pastures for their livestock. (Photograph by Mauricio Antón)

PLATE 7
Semiarid savanna in Samburu National Park, Kenya

Low rainfall in parts of northern Kenya contributes to the creation of a grassland savanna with very sparse tree cover, few shrubs, and extensive grassland. The umbrella thorn common to this type of savanna and the Grevy's zebras provide a quintessential image of the African savanna. (Photograph by Mauricio Antón)

PLATE 8

Arid savanna with grassland and shrub in Etosha National Park, Namibia

Despite the fossil lake bed in the distance, low rainfall and the absence of large bodies of permanent water create a rather hostile landscape in Etosha, where such water-dependent animals as hippos and buffalo are absent. A few wildebeest and springboks appear behind the herd of elephants. (Photograph by Mauricio Antón)

PLATE 9

Life reconstruction of a male *Theropithecus brumpti*

The fossil baboon *Theropithecus brumpti* was strongly sexually dimorphic. The males not only were much larger than the females, but also had impressive canine teeth that were used in displays and actual fights. A pattern of contrasting colors in and around the face is a logical inference for these monkeys, which frequented the shaded environment of riverine woodlands where such features would make visual communication more effective.

PLATE 10

Life reconstruction of females and children of *Homo erectus*

It has been suggested that *Homo erectus* lived in larger, more structured social groups than earlier hominin species. If this is the case, it is probable that ape-like grooming was no longer a practical way of maintaining cohesion among the group members. A new, characteristically human way of strengthening social bonds may have appeared at this stage—laughter.

PLATE 11

Life reconstruction of *Megistotherium osteothlastes*

Although given a separate genus name, it is possible that this fossil hyaenodontid belongs in the genus *Hyainailouros*. It is known mainly on the basis of a complete skull, larger than that of any known creodont or of any extant or extinct member of the order Carnivora, for that matter. Like the other members of the family Hyaenodontidae, however, *Megistotherium osteothlastes* had a very large head relative to its body size, so it is likely that its body mass would not have been larger that of the largest extant bears or big cats. The form of the dentition indicates that this giant creodont was both a hunter and a scavenger.

PLATE 12

Life reconstruction of *Ictitherium ebu*

The long-limbed proportions of the fossil hyena *Ictitherium ebu* point to a developed running ability, but do not imply pursuit hunting of big game, as in modern wolves and spotted hyenas. The small body mass and generalized dentition suggest a diet more like that of modern jackals and coyotes, which includes small to medium-size mammals and carrion.

PLATE 13

Life reconstruction of *Megantereon cultridens*

Although *Megantereon cultridens* was thoroughly cat-like in general appearance, the differences in skull proportions between this machairodont cat and modern cats would have been evident in the living animal. The orbits, for instance, were oriented less frontally. Thus in a view like that shown here, the eye on the far side would not have been visible to the observer, while it is in a leopard or jaguar. In spite of this more lateral orientation of the eyes, *Megantereon* enjoyed very reasonable binocular vision, comparable to that of a wolf.

PLATE 14

Life reconstruction of *Nyanzachoerus pattersoni*

The fossil pigs of the genus *Nyanzachoerus* displayed some of the most spectacularly ornamented skulls of the family Suidae. They were also enormous animals, bigger than the forest hog, which is the largest of the extant suids. An adult male nyanzachoere would have been an impressive sight, and an opponent to be avoided by most, if not all, predators.

PLATE 15

Life reconstruction of *Hippopotamus gorgops*

The fossil hippopotamus *Hippopotamus gorgops* was similar in size and proportions to its modern coun-terpart, the common hippo, but there were interesting differences in the skull. The orbits were in a more elevated (periscopic) position than in the extant hippos, and the shape of the occiput indicates that the snout could be elevated at a greater angle. During yawning displays in the water, the mandible is kept level with the surface, and so the skull must be elevated in order to attain the larger gapes. These features suggest a more extreme adaptation to amphibious life.

PLATE 16

A scene at East Turkana, Kenya, during the early Pleistocene

During the dry season, a number of mammal species characteristic of the Koobi Fora area of East Turkana gather on the banks of a fluvial channel near the point where it joins a paleo-lake: the elephant *Elephas recki*, the reduncine antelope *Kobus sigmoidalis*, the scimitar-toothed cat *Homotherium latidens*, and the zebra *Equus koobiforensis*.

4 FOSSIL SITES IN AFRICA

AFRICA HAS A RICH vertebrate fossil record, but there are inevitable biases. Much of the material, especially that of the past 5 Ma, during which humans have emerged, comes from the eastern side of the continent, although southern and northern regions provide important evidence. Since we have no reason to suppose that animals lived only in those parts of Africa from which their fossil bones are recovered, it is evident that the true picture has been distorted by various factors, three of which may be highlighted. The first is the simple fact that not every animal that dies enters the fossil record; it may be completely eaten or may decay beyond the stage where there is anything left to be preserved. Although we cannot hope to be precise about the number of animals thus lost to the fossil record, such is probably the fate of most carcasses. The second is that suitable deposits for the preservation of fossils may not be accumulating in any particular region. Sedimentation in one place is basically the product of erosion somewhere else; without sedimentation, there is nothing to bury the remains and preserve the fossils—and even such sediments as do accumulate may not preserve bone particularly well. The combination of these two factors means that whole populations of particular species may be absent from the record of certain regions. The third factor is the visibility of fossil-bearing deposits; fossilization involves burial, and unless the deposits are subsequently revealed by natural processes such as erosion, or by quarrying or deliberate exploration, the fossils will not be recovered.

Despite all the problems, we are fortunate in Africa. The massive depositional basins of eastern Africa are filled with sediments eroded from the highlands produced

FIGURE 4.1

The distribution of major fossil localities in Africa

The major fossil localities in Africa: *1*, Fayum Depression, Egypt; *2*, Gebel Zelten, Libya; *3*, Rusinga and Mfwangano Islands, Lake Victoria, Kenya; *4*, Maboko Island, Lake Victoria, Kenya; *5*, Fort Ternan, Kenya; *6*, Arrisdrift, Namibia; *7*, Bled ed Douarah, Tunisia; *8*, Lothagam, Kenya; *9*, Sahabi, Libya; *10*, Lange-baanweg, South Africa; *11*, Aramis, Ethiopia; *12*, Kanapoi, Kenya; *13*, Koro Toro, Bahr el Ghazal, Chad; *14*, Laetoli, Tanzania; *15–17*, Turkana Basin: Omo River, Ethiopia, and East Turkana and West Turkana, Kenya; *18*, Chiwondo Beds, Malawi; *19*, Hadar, Ethiopia; *20*, Bouri, Ethiopia; *21*, Bodo D'Ar, Ethiopia; *22*, Olduvai Gorge, Tanzania; *23*, Taung, South Africa; *24*, Sterkfontein Valley: Sterkfontein, Swartkrans, and Kromdraai, South Africa; *25*, Drimolen Cave, South Africa; *26*, Gondolin, South Africa; *27*, Makapansgat Limeworks, South Africa; *28*, Elandsfontein, South Africa; *29*, Olorgesailie, Kenya; *30*, Ternefine, Algeria; *31*, Kabwe or Broken Hill, Zambia; *32*, Klasies River Mouth, South Africa; *33*, Ahl al Oughlam, Morocco.

by tectonic events over the past few tens of millions of years, and enough animals obviously died in circumstances in which their carcasses could become incorporated into the deposits. Thus the first and second distorting factors have been locally overcome, and fossil accumulation in eastern Africa often has been considerable. But such accumulation would count for nothing unless continued tectonic and erosional processes had ensured that the sediments were then revealed and dissected for us.

What we therefore have is a patchwork of evidence from localities where a combination of the former presence of a species, the preservation of the remains, and the discovery of the fossils provides us with a series of snapshots of the past. For this reason, it is useful to visit some of the most important sites and examine the evidence they provide for past conditions before considering the larger pattern of evolutionary change. It is also worth stressing that conditions at the time of deposition at many sites were often very different from those of the present day. As outlined in chapters 1 and 2, the physical geography and topography of many parts of Africa has changed immensely over the past several million years. Some sites have been raised or lowered by up to 1000 m, and others that were once close to marine shorelines are now far inland. Yet others still exist in the vicinity of massive volcanoes or the remains of equally huge ones that have come and largely gone since the formation of the fossil site. Even Africa itself 20 Ma ago was some 5 degrees of latitude—around 500 km—south of its present position. Evidence for the original conditions is often to be found in the material preserved in the deposits, although in using such evidence we must be careful to avoid circularity in argument, going from fossils to past environment and from past environment to interpretations of lifestyle of the fossil species.

We present the sites in generally chronological order, from earliest to latest, although some of them, such as those in the Chad and Turkana Basins, are grouped together when it makes more sense of the physical geography of the regions in which they are found. The location of the sites is shown in figure 4.1, and we offer reconstructed scenes for some of the more important localities.

Fayum Depression, Egypt

The Fayum Depression, about 80 km southwest of Cairo and on the western side of the Nile in what is formally termed the Jebel Quatrani Formation, provides one of the earliest records of the primates on the African continent. The deposits range in age from around 40 to 31 Ma, the late Eocene to early Oligocene. Considerable collecting work since the 1870s has provided a wealth of biotic information, and intensive collection in the 1960s led to the recognition of at least nine primate species, with the taxonomic diversity being even more marked than that number might imply since

FIGURE 4.2

A scene in the Fayum Depression, Egypt, during the Oligocene

On the branch above two strangely horned arsinotheres of the species *Arsinoitherium zitteli* is a small monkey-like primate, *Apidium*. This primitive primate is placed in the family Parapithecidae.

perhaps five distinct families are represented. A primitive whale was one of the earliest discoveries, augmented by sirenians (dugongs and manatees), and the larger mammals include primitive proboscideans. In the earliest Oligocene part of the sequence, around 50 percent of the known fauna is made up of hyraxes, easily the dominant and most diverse herbivore group. The overall variety of plant and animal remains recovered suggests a coastal floodplain with mangrove swamps, slow-moving streams, and ponds forming—all making up a rich habitat able to support an extensive range of organisms (figure 4.2).

The Fayum is also well known as the type locality of the rhinoceros-like *Arsinoitherium zitteli*, to modern eyes a bizarre animal with a pair of large bony horn cores attached to the skull, although the range in Africa has now extended to include Ethiopia. Material from Turkey, Romania, and even Mongolia has been linked with the Egyptian species but is much smaller and more primitive.

Gebel Zelten, Libya

The early Miocene site of Gebel Zelten, dated to around 22 Ma, lies some 300 km inland from the Mediterranean and around 200 km to the southwest of Sahabi, an early Pliocene locality. It is an elongated, elevated area, or mesa. It was formed on a coastal floodplain and has a rich vertebrate fauna of both marine and terrestrial species, whose bones have been recovered following erosion of the mesa that has produced numerous steep-sided wadis. The large-herbivore fauna—which includes proboscideans, perissodactyls, and ruminant giraffids as well as pigs, creodonts, and one of the few African occurrences of nimravids in the form of *Syrtosmilus syrtensis*—suggests relatively open country. Reconstructions of Gebel Zelten have suggested a northward-flowing river with a belt of forest and a savanna hinterland in a tropical climate, with similarities to the area where the Omo River flows into Lake Turkana in southern Ethiopia.

Rusinga and Mfwangano Islands, Lake Victoria, Kenya

The small islands of Rusinga and Mfwangano lie at the mouth of the Winam Gulf, the northeasterly extension of Lake Victoria that juts into Kenya. Although Lake Victoria itself is not a direct product of rifting, unlike the Western Rift lakes that circle around it to the west, the Winam Gulf and the two islands do lie within the small, northeast–southwest-trending Nyanza Rift. The early Miocene deposits of Rusinga and Mfwangano, dated to around 17 Ma, were formed on alluvial floodplains. Rifting

FIGURE 4.3

A scene at Rusinga Island in Lake Victoria, Kenya, during the early Miocene

A pair of anthracotheres of the species *Masritherium aequatorialis* cross a small stream with sandbanks, while a flying squirrel (*Paranomalurus bishopi*) glides over the stream and a second squirrel climbs up the trunk of a large tree in the left foreground. In the background, a chalicothere (*Chalicotherium rusingense*) walks along the edge of a forest clearing. The environment is largely forested, with tree genera (*Grewia, Cordia, Celtis,* and *Terminalia*) that are common in gallery forests and Zambezian woodlands in modern Africa. A volcano corresponding to the present-day Gwasi Hills is seen in the distance.

at this point was in its early stages, although there was volcanic activity, and runoff from existing highland relief in southwestern Kenya appears to have ensured considerable local sedimentation and possibilities for fossil accumulation. Rusinga, in particular, has proved to be a rich source of fossils.

The fossil flora, together with paleosol evidence, points to trees and shrubs in dry woodland with open areas and wetter forests (figure 4.3). Large mammals include pigs, anthracotheres, chevrotains, rhinos, chalicotheres, and gomphotheres among the herbivores, while the carnivores are represented by creodonts. Primates are present, essentially ape-like animals referred mainly to the genera *Proconsul* and *Rangwapithecus*, and the fauna again supports an interpretation of habitat diversity.

Maboko Island, Lake Victoria, Kenya

The small island of Maboko lies farther into the Winam Gulf of Lake Victoria than Rusinga and Mfwangano. The fauna comes from mid Miocene deposits dated to around 15 Ma and includes an extinct early hominid probably best considered as a member of the genus *Afropithecus*, although the dominant primate in the lower deposits is the extinct monkey *Victoriapithecus maccinnesi*. The environment at the time of deposition, based on a variety of fossil and sedimentological indicators, appears to have included open woodland and swamps, alkaline and freshwater habitats, and seasonal rainfall.

Fort Ternan, Kenya

The mid Miocene site of Fort Ternan, dated to around 14 Ma, is in a building-stone quarry some 40 km east of the Winam Gulf of Lake Victoria and has been investigated since the 1960s. Rifting was still in its early stages when the deposits were laid down, but lava floods were forming the high plateaus of western Kenya and the hilly areas of the center of the country and perhaps introducing rain shadows. The deposits at Fort Ternan contain a rich fauna, some members of which point to immigration from Eurasia, that is dominated among larger mammals by the remains of bovids and giraffids.

There has been quite a lot of discussion about the paleoenvironment implied by the Fort Ternan evidence, based on analyses of vertebrate remains, plant fossils and pollen, mollusk shells, and the stable isotope chemistry of soil carbonates. Some researchers have taken the evidence from the main fossiliferous deposit to suggest a mosaic of grassland, wooded grassland, and grassy woodland with deciduous trees in

FIGURE 4.4

A scene at Fort Ternan, Kenya, during the mid Miocene

The viverrid *Kanuites* balances on a branch, while the smaller bovid *Kipsigiceros* and the larger giraffid *Palaeotragus primaevus* approach the banks of the drainage channel. The environment is considerably drier than in early Miocene times, and arboreal vegetation is seen to concentrate along the channel margins, while farther away the woodlands become more open and there are patches of grassland. The composition of the tree cover probably resembled that of modern-day Zambezian woodland, with an abundance of acacias.

the area. They argue for a generally drier climate than that in the earlier part of the Miocene, although gastropod remains also indicate evergreen forest in the vicinity, and freshwater was obviously present. Other interpretations favor a generally more closed habitat, and much of the argument appears to hinge on whether undoubted grasses represent grasslands or simply highly localized conditions such as patchy areas within an essentially wooded landscape. Certainly there is no evidence for specialized grazers among the ungulate fauna. Deposition appears to have occurred in a shallow gully, presumably a drainage channel from the slopes of nearby Mount Tinderet, concentrating material from the variety of habitats sampled (figure 4.4). Volcanic ash fall from Tinderet has been implicated in the formation of such a large accumulation of well-preserved fossil remains.

Arrisdrift, Namibia

Arrisdrift, a mid Miocene site in the southern part of the Namib Desert in the very south of Namibia, is on the northern bank of the Orange River. The deposit itself appears to lie in a seasonally flooded channel of the proto-Orange (figure 4.5). Many of the fossils have cracked surfaces typical of exposure to the sun while fresh, while others bear teeth marks of crocodiles. The mammalian fauna consists of a range of species adapted to closed and open habitats. They include the small bovid *Namacerus gariepensis*, a chevrotain of the genus *Dorcatherium*, pigs of the genera *Namachoerus* and *Nguruwe*, and a long-legged rhino—all with the low-crowned dentition of browsers. The presence of both a gomphothere and a small early deinothere (*Prodeinotherium hobleyi*) is thought to reinforce the interpretation of largely closed vegetation.

Remains of a lagomorph (*Kenyalagomys*), an elephant shrew (*Myohyrax*), and a springhare (*Parapedetes*) are thought to indicate grassy glades, however. Six species of small to large true carnivores, including the bear-size amphicyonids *Amphicyon giganteus* and *Ysengrinia ginsburgi*, have been identified at Arrisdrift, and a large creodont is also known from the locality. Both the creodonts and the amphicyonids may have specialized in taking members of the larger herbivore species.

Bled ed Douarah, Tunisia

The mid to late Miocene site of Bled ed Douarah, dated to somewhere in the range 14 to 10 Ma, is located some 40 km inland from the Mediterranean. The fossil-rich horizons are referred to the Beglia Formation and seem to have formed close to a delta

FIGURE 4.5

A scene at Arrisdrift, Namibia, during the mid Miocene

In the dry, open woodlands, species of *Terminalia* and *Combretum* are the dominant trees. The fluvial channel, which for most of the year probably is a pool of quiet waters, flowing only at the height of the rainy season, eventually became the depository of the fossils found at the site. Crocodile fossils are very abundant, indicating a warmer climate than exists today in the area, and crocodile tooth marks on several mammalian bones testify to the dangers of drinking in that pool. Among the species found at Arrisdrift are (*from left to right*) the mastodon *Afromastodon coppensi*; the deinothere *Prodeinotherium hobleyi*; the giant hyrax *Prohyrax hendeyi*; the tiny ruminant *Namacerus gariepensis*; the cat-like carnivore *Diamantofelis ferox*; the climacoceratid *Orangemeryx hendeyi*; and the crocodile *Crocodylus gariepensis*.

with an open-country hinterland. Numerous taxa are recorded, including deinotheres, mastodons, anthracotheres, pigs, the caprine bovid *Pachytragus solignaci*, and three-toed horses of the genus *Hipparion* among the mammals as well as freshwater fish and birds indicative of both open-country and freshwater habitats. The material formerly identified as suid is now considered to consist of both a tetraconodont pig of the genus *Nyanzachoerus* and a primitive hippopotamus of the genus *Kenyapotamus*. Carnivores from the Beglia Formation include hyenas of the genera *Proticititherium* and *Lycyaena* and an amphicyonid of the genus *Agnotherium*, as well as a large machairodont referred to *Machairodus* and a nimravid referred to *Sansanosmilus palmidens*.

Lothagam, Kenya

Lothagam lies in the Turkana Basin close to the western shore of Lake Turkana (formerly Lake Rudolf) in northern Kenya. The site has been known since the late 1960s, but publications of studies of the deposits and biota have been sporadic. The main fossil-bearing strata fall in age between 7 and 5 Ma—in other words, the latest Miocene—a period of great importance in the evolution of the African Hominidae and the time at which biomolecular evidence points to the split between the great apes and humans. Unfortunately, although cercopithecid monkeys are known, the only hominid material recovered from the site consists of two isolated teeth and a fragment of mandible, none of which permits clear identification as hominin. But the rest of the mammalian assemblage is rich, including around a dozen carnivores, four proboscideans, three equids, two rhinos, three suids, two giraffes, and 19 bovids. Among the material are three remarkably complete skeletons of carnivores, representing a saber-toothed cat referred to the new genus and species *Lokotunjailurus emageritus*, a new species of hyena of the genus *Ictitherium*, and a giant mustelid. The earliest currently known appearance of the false saber-toothed cat *Dinofelis* is also recorded at Lothagam.

The deposits at the site are thought to have been laid by a meandering river of some size. Abundant fish, oysters, crocodiles, turtles, and hippos—the last the most frequently preserved mammal—point to a perennial flow and large bodies of water. These indicators plus the mammalian fauna in general, including numerous remains of impala as well as of alcelaphine and reduncine antelopes, suggest the presence of swamps, woodland vegetation close by. and grassy, more open plains in the general vicinity (figure 4.6).

Chad Basin, Chad

The two important sites in the Chad Basin are Toros-Menalla, with *Sahelanthropus tchadensis*, and Koro Toro, with *Australopithecus bahrelghazali*. The sites are close together in the locality, some 2500 km west of the Great Rift Valley, within the boundary of the former Lake Chad. The remains attributed to *S. tchadensis* at Toros-Menalla derive from perilacustrine sandstones dated by associated fauna to 7 to 5 Ma, the late Miocene. That fauna—with freshwater fish, hippos, otter, and crocodiles—indicates the presence of a large and permanent body of water at the time of deposition, with swamps and gallery forest and nearby grasslands—in other words, a mosaic environment. Among the mammalian fauna is the anthracothere *Libycosuarus petrocchi*;

FIGURE 4.6

A scene at Lothagam, Kenya, during the late Miocene

In woodlands near the permanent river, a pair of saber-toothed cats of the species *Lokotunjailurus emageritus* pursue three hipparionines (*Eurygnathohippus turkanense*) onto a patch of grassland.

hexaprodont hippos; the giraffe *Sivatherium*; a variety of bovids, including kobs and hippotragines; the pig *Nyanzachoerus syrticus*; hipparion horses; monkeys; smaller hyenas; and a machairodont, possibly *Machairodus giganteus*.

The Chad Basin is the most northern and most western area to have produced remains attributed to the hominin genus *Australopithecus*. The discoverers initially referred the hominin remains from Koro Toro to *Australopithecus afarensis*, but later proposed that they be reassigned to a new taxon: *A. bahrelghazali*. The hominin fossils, together with those of several other vertebrates suggesting an age in the vicinity of 3.4 to 3 Ma, come from deposits indicative of a low-energy fluvial system in the Chad Basin, a conclusion supported by remains of fish, crocodile, and a hexaprotodont hippo (*Hexaprotodon protamphibius*). The vertebrates include ungulates thought to indicate more wooded areas, such as a reduncine antelope (*Kobus*) and a pig (*Kolpochoerus afarensis*), together with an elephant (*Loxodonta exoptata*). A three-toed horse (*Hipparion*), a white rhino (*Ceratotherium* cf. *C. praecox*), and an unidentified alcelaphine antelope represent more open-country species. Overall, the evidence has been taken to indicate a lakeside location with streams and a mosaic of vegetation.

Sahabi, Libya

Deposits in the vicinity of Qasr as-Sahabi, some 200 km inland from the Gulf of Sidra, record the Messinian salinity crisis at the boundary of the Miocene and the Pliocene, about 5 Ma, when the Mediterranean became landlocked and virtually dried up. The main fossil-bearing deposit is of earliest Pliocene age and records the refilling of the Mediterranean, with the lowermost levels containing a marine fauna and the uppermost giving way to terrestrial deposits and channel fills with terrestrial mammals. Deeply eroded channels to the south of the locality—down to some 500 m below sea level—are of an age equivalent to that of the extensive deposits of gypsum, an evaporite mineral (calcium sulfate) derived from marine waters, below the main fossil-bearing sediments. Similarly deep channeling is seen below the present bed of the Nile and, together with the gypsum, indicates the extent of evaporation and the massive fall in sea level after the severing of contact between the Mediterranean and the Atlantic.

The terrestrial components of the Sahabi Formation point to wooded habitats close to the banks of a fairly large river, but the surroundings appear to have been arid or semiarid, perhaps with an extended dry season. The pigs, bovids, giraffes, rhinos, and equids are distinctly open-country animals, while the smaller mammals are extensively represented by gerbils. Many of the species point to close relationships with faunas of Eurasia, while others appear to be endemic to Africa.

Langebaanweg, South Africa

Langebaanweg lies almost at the southernmost tip of Africa, just a little way up the western coast about 100 km north of Cape Town. The extensive fossil remains of numerous vertebrate species lie in phosphate deposits of latest Miocene to earliest Pliocene age, around 5 Ma. The lowermost deposits are of beach gravel with a marine fauna, but terrestrial strata above contain a wide range of land mammals. Although frequently disarticulated, the fossils often consist of complete or semicomplete bones in remarkable states of preservation. The sample of carnivores is one of the richest from any African open-air deposit, with several species of hyenas and cats, while the range of ungulates is extensive. Grasslands and riverside woodlands close to a coastal area are indicated at Langebaanweg.

Aramis, Ethiopia

The Aramis drainage area is in the Middle Awash Valley of Ethiopia, to the south of Hadar. The important deposits occur just above a volcanic tuff dated to around 4.4 Ma, the early Pliocene. The locality has produced remains of the hominin *Ardipithecus ramidus*, the most primitive known member of the Homininae.

A rich mammalian fauna that includes tragelaphine antelopes and colobine monkeys, together with plant remains, has been taken to indicate woodland conditions in the vicinity, although the extent of such cover is unclear. The habitat preferences of the monkeys have not been established, nor is it evident that the tragelaphines were confined to woodlands. The fauna might indicate extensive forest or localized gallery forest along a watercourse in an area of more open terrain. The bones of all the medium-size and large mammals, including the hominins, show evidence of carnivore activity and no indication of water transportation, suggesting a flat plain with scattered carcasses as the likely setting. The pollen remains indicate that the region has been downfaulted by perhaps 1000 m since 2.9 Ma, because the vegetation is typical of more montane areas.

To the southwest of Aramis, but still in the Middle Awash Valley, is a further series of downfaulted localities—such as Saitune Dora, Alayla, Asa Koma, and Digiba Dora—where hominin remains referred to the subspecies *Ardipithecus ramidus kadabba* have been recovered from deposits bracketed to an age of 5.8 to 5.2 Ma. The associated habitats have been interpreted as higher-altitude woodland to grassy woodland on the basis of soil carbonates and the associated fauna, leading to arguments that ancestral hominins at this time were still very much woodland animals.

Kanapoi, Kenya

The early Pliocene site of Kanapoi lies to the south of Lake Turkana. Recently discovered hominin material in well-dated deposits of around 4.1 Ma has been referred to the species *Australopithecus anamensis*.

Deposition appears to have been water-laid in what has been interpreted as a precursor of the present-day Lake Turkana. More than 30 mammalian taxa are recorded from the site, together with remains of fish and aquatic reptiles, implying a range of habitats in the vicinity. As at Aramis, carnivores damaged many of the bones, including those of the hominins.

Laetoli, Tanzania

Laetoli is located to the south of Olduvai Gorge, between Olduvai and Lake Eyasi. The fossil-bearing deposits fall in the age range 3.8 to 3 Ma. The site is unique for two main reasons. First, the deposits there, unlike those at other eastern African sites, accumulated in dry savanna away from any water source and therefore throw a potentially different light on habitats and environments of the mid Pliocene. The range of plants and animals represented indicates a mixture of grass- and woodland habitats. Second, the large assemblage of mammalian species and numerous remains of a hominin referred to the species *Australopithecus afarensis* are supplemented by a collection of trackways. They formed in wet volcanic ash, which hardened to preserve a set of footprints (figure 4.7), and include the hoofprints of numerous ungulates and the impressions of a bipedal creature inferred to have been *A. afarensis*. We therefore have the opportunity to compare the skeletal characters of this species and our inferences about the functional anatomy underpinning its bipedal locomotion with the direct effects of such a movement pattern.

Turkana Basin
Omo River, Ethiopia

The Omo River, which rises in the highlands of Ethiopia, runs south into Lake Turkana in Kenya, where deposits on the eastern and western shores are informally

FIGURE 4.7

A scene at Laetoli, Tanzania, during the mid Pliocene

The lower unit of the Footprint Tuff at Laetoli appears to have accumulated at the end of the dry season, and the presence of tracks of dik-dik, guinea fowl, rhinoceros, and giraffe is consistent with a population of resident animals that remain in the area year-round. Footprints of equids, proboscideans, and other more water-dependent species appear only in the upper unit, suggesting the arrival of the rains in the dry savanna.

known as East and West Turkana, respectively. In formal terms, the deposits of the Omo River, together with those of East and West Turkana, are referred to as the Omo Group within the Turkana Basin. In the Omo River region, the important fossil-bearing elements of the Omo Group are the Usno and Shungura Formations, the Koobi Fora Formation at East Turkana, and the Nachukui Formation at West Turkana. Craig Feibel and Frank Brown have undertaken considerable work on the dating and the stratigraphic relationship of these deposits. Material is now also appearing from deposits associated with the Turkwell River, which runs into the western side of the lake, but details have not yet been published.

Events within the Turkana Basin are reflected in the accumulation of large amounts of sediment. The basin lies at an altitude of 470 m, surrounded by relatively low relief, and most of the water enters Lake Turkana from the Omo River. Today the region is arid, with scrub and annual grasses, but the river appears to have acted as a narrow, elongated oasis of dense vegetation—virtually forest—in a drier landscape of grasslands that from time to time may have been flooded (figure 4.8). Lake Turkana itself is a relatively recent body of water. Over the past 4.5 Ma, others have come and gone, with associated effects on local habitats, and the overall pattern could be described as shifting between large meandering river, at times with braided streams, and lake.

The Shungura Formation, totaling some 765 m in thickness, consists of 12 deposits or members, each marked by a volcanic tuff at its base. Numerous other tuffs exist within the deposits, but not all offer basinwide correlations. The Shungura members range in age from 3.5 to nearly 1 Ma, and have provided a framework of absolutely dated volcanic markers and chronologically well constrained fossil faunas against which other volcanic markers can be matched. Other assemblages can be fit into that framework for the latter part of the Pliocene and the earliest part of the Pleistocene. This is a crucial period in the evolution of the African biota and, particularly, the human lineage.

East Turkana, Kenya

The Koobi Fora Formation at East Turkana, which takes its name from the prominent spit of land that juts into the eastern side of Lake Turkana, has deposits totaling about 550 m thick. They are much the same age as those of the Shungura Formation of the Omo River region, and consist of eight members again interspersed with volcanic tuffs that provide the basis for dating and correlation. More than 80 species of small mammals and over 90 species of larger mammals are known from East Turkana, together with fish, sponges, mollusks, and extensive floras of diatoms and of higher plants based on pollen and plant macrofossils (plate 16).

FIGURE 4.8

A scene at the Omo River, Ethiopia, during the late Pliocene

During the formation of Member G of the Shungura Formation, at about 2.3 Ma, the assemblage of mammals includes (*from back to front*) the deinothere *Deinotherium bozasi*, the bovid *Simatherium shungurese*, the impala *Aepyceros melampus*, and the baboon *Theropithecus brumpti*. The environment corresponds to a riverine setting with a grassy floodplain, a gallery woodland on slightly higher ground, and spectacular palms of the genus *Borassus*, which are recorded in the Shungura Formation since Pliocene times.

The locality of Allia Bay has produced some of the oldest known hominin fossils, with an age of around 3.9 Ma.

West Turkana, Kenya

The deposits of West Turkana are referred to the Nachukui Formation, which is divided into eight members with volcanic tuffs. The semicomplete skeleton of a juvenile *Homo erectus* (KNM-WT 15000) comes from these deposits, in the lowest part of the Natoo Member, with a date of 1.53 Ma (figure 4.9). From lower down in the same sequence comes the type specimen KNM-WT 40000, now referred to one of the newest hominin taxa, *Kenyanthropus platyops*, with an interpolated age estimate of 3.5 Ma. In all, nearly 100 species of larger mammals are known from West Turkana, a large proportion of which are antelopes.

Chiwondo Beds, Malawi

The Pliocene lake deposits of the Chiwondo Beds, on the northwestern shore of Lake Malawi, have provided a much-sought intermediate fossil locality between those of eastern and southern Africa. Lake Malawi sits squarely within the western arm of the Great Rift Valley and is the product of the rifting process. Although the deposits have been known since the 1920s, work at the beds is still at an early stage, and material collected so far has come from thin surface scatters across fairly large areas. The age of the remains is so far unconstrained, but it is considered to be between 4 and 1.6 Ma.

A wide range of species has been found, with bovids and pigs prominent among a mammalian fauna that includes proboscideans, horses, rhinos, hippos, giraffes, and a camel in addition to a hominin, represented by a mandible. Fish, turtles, crocodiles, and mollusks also have been recovered, and the whole assemblage points to a mosaic of closed, wet and dry environments in the vicinity of a small lake.

Hadar, Ethiopia

Hadar lies in the Afar Depression close to the Awash River. The depression, the northernmost part of the Great Rift Valley on the African continent, is a sedimentary basin with a great depth of deposits and a number of fossil-bearing members interbedded with volcanic layers (figure 4.10). As at the nearby locality of Aramis, the deposits at Hadar have been downfaulted by around 1000 m since deposition.

The Hadar site has given the name to the hominin species *Australopithecus afaren-*

FIGURE 4.9

A scene at West Turkana, Kenya, during the early Pleistocene

During the formation of the Natoo Member of the Nachukui Formation, *Homo erectus* (or *H. ergaster*, according to some) creeps up on three kobs (*Kobus kob*) and a group of bovines of the species *Pelorovis olduwayensis*. The edaphic grasslands provide an ideal environment for kobs, and the riverine woods contain tree species recorded as fossils from the Omo Group deposits, including the fig tree *Ficus* and palms of the genus *Hyphaene*.

FIGURE 4.10

A scene at Hadar, Ethiopia, during the mid Pliocene

One of the streams coming from the nearby highlands forms a small delta as it joins the lake in the basin bottom. Some reduncine antelopes of the genus *Kobus* graze the floodplain grasses, while two giraffes of the species *Giraffa jumae* browse among the gallery woods. The proboscideans are represented by a trio of elephants (*Elephas recki*) and, in the distance, two deinotheres of the species *Deinotherium bozasi*.

sis, although the sample on which the name is based includes material from Laetoli, from which the type specimen, a lower jaw, actually comes. The age of the hominin fossil material is around 3 Ma, based on the dating of the volcanic deposits of the Sidi Hakoma Tuff.

Ahl al Oughlam, Morocco

Ahl al Oughlam, with deposits dated to around 2.5 Ma, is one of the richest localities of the late Pliocene in northern Africa. Dozens of species have been recorded, with 23 of carnivores alone. The remains were recovered from sediment-filled fissures, and these predators appear to have played a considerable part in the accumulation.

Bouri, Ethiopia

Bouri is the name given to the locality in the Middle Awash Valley, about 30 km east of Aramis, where the recently discovered hominin material assigned to *Australopithecus garhi* was found. The specimens actually come from the Hata Member of the Bouri Formation, dated to 2.5 Ma, which also has produced remains of other primates as well as vertebrates that include 19 antelope species, hippos, giraffes, pigs, equids, proboscideans, and carnivores. The structure of the herbivore fauna, with an abundance of such grazing species as alcelaphine and antilopine antelopes as well as species dependent on water, and the nature of the sediments point to a habitat on the margin of what is believed to have been a shallow freshwater lake.

Several of the mammal bones bear marks interpreted as cutting and breakage, believed to indicate disarticulation, defleshing, and marrow retrieval, and would constitute the earliest evidence for such behavior by hominins.

Olduvai Gorge, Tanzania

Olduvai Gorge must rank as perhaps the most famous site of exploration and discovery in the search for human ancestry, based on the material recovered from it and its association with the Leakey family since the 1930s. Deposits at Olduvai range in age from around 2 Ma onward. The remains include the type material of *Homo habilis* and the robust australopithecine now referred to *Paranthropus boisei,* but formally referred to the genus *Zinjanthropus* and known informally as nutcracker man because of the enormous teeth and chewing apparatus.

The gorge is an eroded section, up to 90 m deep and around 45 km long, cut into the bed of a former saline lake on the eastern edge of the Serengeti Plain (figure 4.11), a present-day mecca for tourists eager to see game. Layers of volcanic tuff interbed the deposits, and toward the base of Bed I is a layer of basalt. It was this basalt that gave the first indication of the true time span of human evolution when it provided a date of 1.7 Ma, the early Pleistocene, in the early 1960s.

Taung, South Africa

The site of Taung has an importance out of all proportion to its size, fossil preservation or richness, or even visibility. It was from there that our first known African fossil relative was recovered and described by the anatomist Raymond Dart in 1925, but the precise site no longer exists. The deposit formed in a cave in tufa, a very pure form of limestone precipitated from mineral-rich springwater, that has long since been quarried away for the building industry. Indeed, the fossils were found only through quarrying, when the workers kept what appeared to be mainly a collection of baboon skulls.

The hominin fossil, the incomplete skull of an individual about five years old at death, was recognized as something rather different from baboon remains by Dart, then the Professor of Anatomy at the University of the Witwatersrand in Johannesburg. He named it *Australopithecus africanus*, or the southern ape-man from Africa, based largely on the form of the dentition and the relative size of the brain. An interesting recent study of the material by the anthropologist Ron Clarke has suggested that the bone accumulation may have been made at least in part by the predatory activities of eagles (figure 4.12).

Sterkfontein Valley, South Africa

The Sterkfontein Valley, actually the valley of the Bloubank River, lies to the northwest of Johannesburg and contains the three famous hominin-bearing localities of Sterkfontein, Swartkrans, and Kromdraai. They are located within sight of one another: Sterkfontein and Swartkrans face each other across some 500 m of the valley, while Kromdraai is about 2 km to the east. The sites consist of former caves in dolomitic limestone, although erosion has reduced them all to fossiliferous deposits in what are now effectively holes in the bedrock.

Deposits at the sites range in age from perhaps 3.5 Ma, the mid Pliocene, until relatively recent times. No absolute dates are currently available, so dating depends on

FIGURE 4.11

A scene at Olduvai Gorge, Tanzania, during the early Pleistocene

A small group of lions (*Panthera leo*) steers clear of a herd of bovines of the species *Pelorovis oldowayensis* that approaches the shore of the paleo-lake for a wallow in the mud. The saline lake is surrounded by a grass-covered floodplain, in turn fringed by a band of woodland, where giraffes of the species *Giraffa jumae* browse from the acacias and other trees.

FIGURE 4.12

A scene at Taung, South Africa, during the late Pliocene

A crowned eagle (*Stephanoeatus coronatus*) carries the body of the young *Australopithecus africanus*.

correlation of the faunas with those from well-dated deposits in eastern Africa. The earliest deposits are those of Member 2 at Sterkfontein, but the main hominin-bearing deposit is Member 4, dated to around 2.8 to 2.4 Ma, from which numerous remains of *Australopithecus africanus* have been recovered (figure 4.13). Deposits at Swartkrans date from around 1.6 Ma and have produced equally numerous remains of *Paranthropus robustus* as well as a smaller quantity of material assigned to the genus *Homo*. Kromdraai, although rich in mammalian fossils, especially those of larger carnivores (figure 4.14), has yielded few hominins, but the type material of *P. robustus* is from this locality, with a probable age of around 1.6 Ma.

All three sites have produced extensive assemblages of larger mammals, with an impressive diversity of antelope species (figure 4.15) and a full range of carnivores that include hyenas, saber-toothed cats, and such living cats as the lion and leopard. Although stone tools are known from Sterkfontein and Swartkrans, implying activity at the localities, it has long been recognized from the work of C. K. Brain at Swartkrans that perhaps the majority of the animal bone, including that of the hominins, probably is the result of predatory activity by carnivores.

Study of the ungulates from the Sterkfontein Valley by Elisabeth Vrba led her to recognize an important transition in the structure of the fauna between perhaps 2.5 and 2 Ma, with the tribal composition of the antelopes, in particular, pointing to a

---

FIGURE 4.13

A scene at Sterkfontein, South Africa, during the late Pliocene

Sterkfontein is one of several African sites where the remains of hominins, such as *Australopithecus africanus*, are found with those of the "false" saber-toothed cat *Dinofelis barlowi*. It has been suggested that some caves were used as dens by these felids, which would have carried back the carcasses of hominins and other prey animals, a suggestion that conjured the picture of *Dinofelis* as a specialist hunter of hominins and other primates, like baboons. It is obvious that *Dinofelis* was perfectly equipped to capture and kill such primate prey on occasion, as shown in the illustration, and the cat would have been justifiably feared by our ancestors, but it is unlikely that the primates were its main or preferred prey. Primates make up a small portion of the total biomass in an ecosystem, and, like all modern large carnivores, *Dinofelis* likely would have concentrated its efforts on the more abundant ungulates. The black-backed jackal (*Canis mesomelas*) was already present at Sterkfontein, and it no doubt would have hung around at a prudent distance, in hopes of feeding on the leftovers of a *Dinofelis* kill.

FIGURE 4.14

A scene at Kromdraai, South Africa, during the early Pleistocene

The site of Kromdraai A, also called the Kromdraai "faunal" site, contains abundant remains of open-terrain ungulates, such as equids and alcelaphine antelopes. It also has yielded remarkable material of one individual of the saber-toothed cat *Megantereon cultridens*. Other mammals that live in wooded habitats, such as the bushbuck (*Tragelaphus scriptus*), are also represented by low numbers of fossils. Such a distribution suggests the presence of more or less open, grassland savanna, flanked by belts of more closed vegetation in favorable places, such as along watercourses. It is likely that *Megantereon* would have preferred such closed habitats, stalking the resident prey but now and then venturing to the more open areas to take some of the abundant ungulates dwelling there.

FIGURE 4.15

A scene at Swartkrans, South Africa, during the early Pleistocene

During the formation of Member 1, around 1.8 Ma, alcelaphine antelopes graze in a relatively open landscape, a valley bottom in the highveld, or hilly country, where the various Sterkfontein Valley sites are
located. The bovid assemblage at Swartkrans Member 1 is dominated by such alcelaphine antelopes as
wildebeests of the genus *Connochaetes*, in the foreground, and the damaliscus-like *Parmularius*. Such a
species composition is regarded as an indication of abundant grasslands and rather thin woods in the valley at that time.

change in vegetational composition and structure. She suggested that this transition aligned with a similar shift in the tribal composition of the eastern African antelopes, indicating a response to changes in vegetation on a continent-wide scale. From this initial insight, she went on to derive her hypothesis of the "turnover pulse."

Drimolen Cave, South Africa

Drimolen Cave, some 7 km north of the Sterkfontein Valley localities, is one of the most recently discovered localities to produce early hominin remains in South Africa, with around 80 specimens recorded. Like the Sterkfontein Valley sites, the cave appears to have begun as a sinkhole, and the bulk of the fossiliferous infill has an estimated age of 2 to 1.5 Ma, the late Pliocene to early Pleistocene. The hominin, identified as *Paranthropus robustus*, is associated with a range of other mammalian species that includes baboons; reduncine, antilopine, and alcelaphine antelopes; *Dinofelis* (perhaps *Dinofelis piveteaui*) and another, unidentified large felid; and a hyena—a grouping thought to be indicative of a mixed habitat including open grassland (figure 4.16).

Gondolin, South Africa

The fossiliferous locality of Gondolin, around 35 km northwest of Pretoria, has been known for a quarter of a century, but only in recent years has interest in it been renewed with the discovery of two broken hominin teeth tentatively identified as belonging to the genus *Paranthropus*. The deposit appears to be essentially a wash of material into a collapsed hole, with an age in the region of 2 to 1.5 Ma, suggested by correlation of the mammalian species present. The analysis of the assemblage presented by Virginia Watson clearly points to the accumulation of prey remains by leopards. The most common of the 27 species at the site (a minimum of 33 individuals) is reedbuck (*Redunca arundinum*), an antelope that favors tall grass or reeds and open water.

Makapansgat Limeworks, South Africa

Makapansgat Limeworks is the name given to a cave, or system of caves, with depositional basins. It lies in the Makapan Valley, 20 km northeast of the town of Potgeitersrust in the northern Transvaal. Extensive excavations during the 1930s for

FIGURE 4.16

A scene at Drimolen Cave, South Africa, during the late Pliocene or early Pleistocene

A group of "robust" australopithecines (*Paranthropus robustus*) forages for roots and tubers while a large adult male makes ready to defend them from an approaching *Megantereon cultridens*. Baboons (*Papio robinsoni*) and springboks (*Antidorcas recki*) walk by in the middle distance. The environment corresponds to a lightly wooded hillside with patches of grassland.

travertine, a pure source of lime for the building industry, produced large quantities of vertebrate remains in hard, cemented deposits. They were dumped on the site by the lime miners, and most of the investigations in more recent years have involved efforts to understand the basic stratigraphy of the deposits and to assign the dumped material to the correct horizon within a five-member scheme. Specimens referred to the hominin species *Australopithecus africanus* have been found in association with a diverse mammalian fauna of late Pliocene appearance, and in the absence of absolute dating materials this fauna has been used to assign the deposits of Member 3 to around 3 Ma.

The extensive and heavily broken vertebrate assemblage was used by the anatomist Raymond Dart, identifier and original describer of the Taung juvenile *Australopithecus africanus*, to argue that the australopithecines at Makapansgat had employed a diverse tool kit using animal bones, teeth, and horns—what he termed the "osteodontokeratic" culture. More recent studies, especially by C. K. Brain, the excavator of Swartkrans, have demonstrated convincingly that Dart's arguments were based simply on the superficial appearances of the remains. What he took to be deliberate fash-

FIGURE 4.17

A scene at Makapansgat Limeworks, South Africa, during the late Pliocene

A pair of striped hyenas (*Hyaena hyaena*) add to the assemblage of bones that eventually would be misinterpreted as evidence of the "osteodontokeratic" culture.

ioning is better explained as natural processes of breakage by agents such as larger predators like striped hyenas, known in some numbers from the site (figure 4.17).

Analyses of pollen from the deposits that yielded the hominin material show a variable habitat, with arboreal pollen pointing to an increased level of tree cover at times and, in particular, wet-forest conditions during Member 3 times. These results differ in details from others based on large and small mammals and analyses of sediments, but all seem to indicate a reasonable amount of cover and relatively warm and wet conditions.

Elandsfontein, South Africa

Elandsfontein, located around 100 km north of Cape Town, is an open-air locality of at least mid Pleistocene age. Human presence and activity is represented by stone artifacts and a fragmentary cranium and mandible of generally primitive appearance, similar in observable features to the more complete specimen from Kabwe, or Broken Hill, in Zambia. Other species, particularly brown hyenas, appear to have been active at the site, using it as a denning area, and the extent to which humans were responsible for the patterns of bone debris is open to serious question. The presence of the saber-toothed cat *Megantereon* suggests an even earlier date for at least part of the assemblage.

Olorgesailie, Kenya

Olorgesaille, an open-air complex of sites, lies in former lake beds in the Great Rift Valley of southern Kenya, to the northeast of Lake Magadi and about 50 km southwest of Nairobi. It is famous for producing thousands of Acheulean hand axes as well as animal bones, particularly those of the large baboon *Theropithecus oswaldi*. Whether the bones and tools represent human hunting, human scavenging, or a combination of processes including hydraulics that have simply brought them together is unclear. The various layers discerned have dates in the region of 900,000 to 700,000 years ago, based on potassium–argon dating of volcanic deposits.

Bodo D'Ar, Ethiopia

In the Middle Awash Valley is the site of Bodo D'Ar, some 50 km north of Bouri. Deposits here are currently estimated to be around 600,000 years old. The partial hominin cranium from Bodo has been various regarded as *Homo erectus*, *H. heidelbergensis*, or an archaic version of *H. sapiens*.

Ternefine, Algeria

The locality of Ternefine, some 20 km to the east of the town of Mascara—famous for giving its name to a cosmetic product—is dated to the earlier part of the mid Pleistocene, perhaps around 600,000 years ago. The deposits appear to have formed in a small lake set in a rather open habitat described as steppe or savanna, and are dominated by the alcelaphine antelope *Parmularius ambiguus*. The site is most noted for producing a primitive hand-ax industry and a series of mandibles, the earliest known North African hominin fossils, generally referred to *Homo erectus*.

Kabwe or Broken Hill, Zambia

The name of this later mid Pleistocene locality, dated to around 200,000 years ago, may cause some confusion. The name given to the site on discovery was Broken Hill, a cave filled with fossil material uncovered during lead and zinc mining in what was then Northern Rhodesia. The virtually complete human cranium found in association with the remains of lions, leopards, elephants, zebras, rhinos, and smaller species was therefore known as the Broken Hill skull or, more often, the Rhodesian skull. Political changes have removed Northern Rhodesia from the map, and the Zambian name Kabwe has replaced Broken Hill. But the designation Rhodesian skull lingers on, since the specimen was used in the formal classification of a hominin species (*Homo rhodesiensis*) or of a subspecies (*H. sapiens rhodesiensis*) of modern humans. More recent taxonomies would classify it as *H. heidelbergensis*.

Klasies River Mouth, South Africa

The Klasies River Mouth, a cave site on the southeastern coast of South Africa, has produced a small number of remains of early anatomically modern humans dated to around 150,000 years ago. Its principal significance, however, lies perhaps in the debate about the implications of the patterning on animal remains recovered from the deposits for interpretations of the hunting abilities of early modern humans. The patternings probably say more about the recovery and analytical methods employed than they do about ancient human activity.

5 The Evolving African Mammalian Fauna

THE EVOLUTION OF the African mammals over the past 35 Ma or so occurred against the background of changes in the climate and physical environment of the continent. The major events during the period included the immigration and emigration of taxa, the turnover and extinction of species, the appearance of the modern fauna, and the evolution of humans. With so many species coming and going over such a long time during which significant changes in the physical environment took place, a number of problems beset any attempt at a synthesis.

Difficulties of Synthesis

We may talk of the "African mammalian fauna," but Africa is a large, environmentally diverse continent and few large mammal species are naturally ubiquitous. Regional and local habitat variation (or mosaicism) undoubtedly played a part in determining past distributions, as it does today. To add to the problem of presenting an overview of the evolution of the mammals of Africa, there is a patchy fossil record, with some parts of the continent better represented than others. There are also frustrating chronological gaps and periods of poor resolution. The early Miocene and the earlier part of the mid Miocene, from 23.5 to 15 Ma, are reasonably well represented, as are most of the Pliocene and the Pleistocene, the past 5 Ma. The second half of the Miocene is less well documented, however, despite a continuing program of work. What remains clear is that faunal evolution in both structure and content was occurring throughout the latter part of the Miocene, and that immigration and extinction—in addition to in situ evolution of new taxa—were important factors. The material from the site of Lothagam in Kenya is of some significance in the discernment of this pattern and the transition from forest ecosystems to the more open, mosaic savanna habitats that characterized much of the Pliocene and Pleistocene of eastern and southern central Africa.

To add to our difficulties, the present-day pattern of habitat variation has itself evolved as the physical geography and climate of Africa have changed. In other words, the evolution of the biota that we seek to understand has taken place against a constantly shifting background. The oscillations in global climate over the past few million years have occurred over a relatively short timescale in geological and evolutionary terms, and we should not expect evolution in terms of speciations and extinctions to track relatively short-term swings in climate. Africa has an almost unique distribution of land across the equator, and much of the continent contains a mosaic-like pattern of habitat distribution. Short-term cyclical swings in climate are likely to have induced similarly cyclical changes in the distribution of habitats and of the component members of ecological communities, without necessarily inducing either speciation or extinction—that is to say, without inducing faunal turnover. Instead, we should expect correlations between *significant steps* in environmental variables and evolutionary changes in narrow versus widely adapted lineages.

Oligo-Miocene: Africa Joins Eurasia

What, then, can we say? Africa today has the greatest diversity of ungulates—particularly antelopes—and the greatest diversity of large carnivores found anywhere on Earth. They are such a well-known feature of the tourist vision of the savannas that it is difficult to picture the continent without them, and yet none of the living species is known until 5 Ma.

Around 30 Million Years Ago

Around 30 Ma, in the late Oligocene, there were no antelopes, nor were there any horses, rhinos, or pigs in Africa. Although there were artiodactyls in the form of the pig- or hippo-like anthracotheres, the dominant medium-size herbivores were hyraxes, based on diversity, size, and ecomorphology. Their relatives in the Afrotheria, the proboscideans of the time, looked little like the elephants of today, although they were moderately large and had the typical pillar-like limbs of their modern descendants. The large *Arsinoitherium*, with its odd, double horn, may have looked like a rhinoceros but is likely to have had a very different, probably partly aquatic, lifestyle. The carnivores were made up of a bizarre collection of animals that did not even belong to the order Carnivora but to the Creodonta. Of all the mammals, the primates of the time were perhaps most like their living descendants in very general appearance, although they seem to have been members of now extinct families.

These animals lived in an Africa that appears to have been broadly tectonically stable, with no eastern and southern areas of uplift, no Sahara Desert, and little of the vegetational zonation that would result from the development of a generalized horseshoe of increasingly drier country around the equatorial rain forests of the western central parts of the continent. Indeed, much of Africa may have been covered in lowland tropical–temperate rain forest interspersed with somewhat drier areas of woodland and scrub. But the distribution of landmasses and the climate were changing, and such a preponderance of rain forest was increasingly interrupted in the period leading up to around 20 to 18 Ma, when the major land-bridge connection between Africa and Eurasia appears to have been established. The seaway to the north was reduced and that to the northeast closed, so the availability of moisture changed. Eastern Africa was beginning to be affected by volcanism and rifting by 30 Ma, and as separation and volcanic outflow proceeded, uplift was adding to the effect of these processes on the physical and biotic environment. The most noticeable impact of uplifting was in the high plateau areas of eastern Africa, where a rain shadow reduced moisture. On a more global scale, the continued closure of the Tethys Sea had the double effect of reducing the evaporative transport of moisture and increasing the distances of many landmasses from oceanic influences. The planet was cooling, and the massive ice cap of Antarctica was beginning to form on the southernmost continent.

Around 20 Million Years Ago

Immigrants from Eurasia at the time of contact with Africa included some odd-toed ungulates, or perissodactyls. But the only members of this order that would be recognizable to the modern observer were the earliest rhinos; the rest were made up of the strange-looking chalicotheres, with their horse-like heads and clawed feet. More artiodactyls also made their appearance around 20 Ma, in the early Miocene, with the incursion of the giraffoide climactocerids and the first antelopes as well as the primitive pigs of the genus *Nguruwe*, which are known from Namibia and Kenya at around 17.5 Ma and therefore must have traversed the continent (figure 5.1). Cercopithecoid monkeys replaced the earlier primates. But contact with Eurasia produced a two-way

FIGURE 5.1

A selection of ungulates from the early Miocene of eastern Africa

(*From left to right*) The climacoceratid *Climacoceras africanus*, the giraffid *Palaeotragus primaevus*, the suid *Kubanochoerus massai*, the rhinoceros *Dicerorhinus leakeyi*, the chevrotain *Dorcatherium crassum*, and the rhinoceros *Brachypotherium heinzelini*.

Each square measures 1 m on a side.

bridge; the proboscideans emigrated from Africa for the first time, as did the hominoid primates. The subsequent history of some of the proboscidean genera during the Miocene suggests a good deal of interchange with Eurasia.

True carnivores entered the continent with the first appearance of the cats and the amphicyonid bear-dogs as well as the mustelids, although the creodonts continued to prosper as the dominant meat eaters and were joined by the cat-like nimravids (figure 5.2). The structure of the guild of larger carnivores seems to have divided into flesh eaters among the early cats, nimravids, and perhaps smaller creodonts, and bone crunchers among the amphicyonids and larger creodonts, some of which, like the widely distributed *Hyainailouros sulzeri*, were enormous animals by modern standards. That both conical-toothed feline and moderately saber-toothed machairodontine cats are known from this period in Africa, while insectivore/omnivore morphologies of the kind seen in mustelids—among which a variety of morphotypes are evident—were found among the earlier stenoplesictids, or false hyenas, serves to underline further the complexity in guild structure. Such complexity points, by implication, to an equivalent complexity in the ecological relationships of predators and prey in the early Miocene.

We can make a number of observations about the wider structural implications of these faunal elements. The ungulates, essentially the prey species, do not appear to

FIGURE 5.2

A selection of creodonts and carnivores from the early Miocene of eastern Africa

(*From left to right*) The larger hyaenodontid *Hyainailouros sulzeri*, the amphicyonid bear-dog *Cynelos euryodon*, the felid or, perhaps, nimravid *Afrosmilus africanus*, and the smaller hyaenodontid *Hyainailouros napakensis*.
Each square measures 1 m on a side.

have been particularly well adapted for speed. The predators do not look as though they were equipped to chase anything moving particularly fast—none of the cats or cat-like members of other groups in any way resembled a cheetah, for instance—so a high-speed chase was clearly not possible. In the absence of the true dogs, it is difficult to assess the extent to which pack hunting was utilized as a predation technique. Dog-like animals, in the form of some of the smaller hyenas of the genera *Ictitherium*, *Hyaenictitherium*, *Lycyaena*, and *Hyaenictis*, appear only much later in the Miocene. The larger amphicyonids are unlikely to have acted cooperatively and probably would have taken a mixture of carrion and hunted meat. These observations are in keeping with interpretations of the earlier Miocene vegetation as generally closed, a habitat in which neither pack hunts nor shorter, high-speed chases offer significant advantages.

Around 15 Million Years Ago

By the mid Miocene, around 15 Ma, conditions in Africa had begun to change further. The closure of the Tethys Sea combined with the effects of uplift to alter the climate and rainfall pattern of the continent, and drier and more seasonal patterns of weather began to have a major impact on vegetation. Fragmentation of the forests continued,

FIGURE 5.3

A selection of ungulates from the mid Miocene of eastern Africa

(*From left to right*) The rhinoceros *Paradiceros mukirii*, the chevrotain *Dorcatherium chappuisi*, the giraffid *Palaeotragus primaevus*, the pig *Listriodon akatikubas*, the climacoceratid *Climacoceras gentryi*, and the bovids *Kipsigicerus labidotus* and *Oioceros tanyceras*.
Each square measures 1 m on a side.

and grasslands began to appear as part of what was clearly a complex mosaic of vegetational types. Such a mosaic must be borne in mind in any effort to interpret changes in vegetation; although the long-term trend was clearly toward an overall reduction in the extent of woodland, local cover still may have been considerable. Nevertheless, more open terrain was appearing, with woodland savannas on the interior plateau of southern Africa and grassed areas in the eastern parts of the continent.

The fauna continued to evolve with fresh waves of immigrations. Hyenas had put in an appearance—although far removed in size, morphology, or behavior from those of today—with the earliest species being civet-like (*Protictitherium*) and mongoose-like (*Proteles* lineage) insectivore/omnivores. As such, they would have augmented the range of morphotypes previously represented by the stenoplesictids and earliest percrocutids. During the mid Miocene, the diversity of the African creodonts was one of the most extensive undergone by the order. True elephants and hippopotamuses evolved, and three-toed hipparionine horses, many of them well adapted skeletally for running in open habitats and for grazing with high, long-wearing teeth, invaded around 10 Ma. So, too, did the tetraconodont pigs of the genus *Nyanzachoerus*, replacing such earlier suid genera as *Listriodon* and *Libyochoerus* (figures 5.3 and 5.4). It is

FIGURE 5.4

A selection of ungulates from the late Miocene of eastern Africa

(*From left to right*) The hippopotamus *Hexaprotodon harvardii*, the pig *Nyanzachoerus syrticus*, the giraffid *Palaeotragus germaini*, the hipparionine *Eurygnathohippus turkanense*, the impala *Aepyceros premelampus*, the tragelaphine antelope *Tragelaphus kyaloae*, and the rhinoceros *Brachypotherium lewisi*. Each square measures 1 m on a side.

from this time onward that the artiodactyls—above all, the antelopes—began to be one of the dominant components of the larger-mammalian faunas, as new immigrants from Asia mixed with a smaller number of what may by then have been indigenous African forms to swell the overall diversity of the ungulates.

Around 8 Million Years Ago

The late Miocene and early Pliocene, the period from around 8 Ma, was a time of considerable further change in Africa. Thure Cerling and colleagues have shown that there was a global pattern of expansion in the biomass of C4 plants in tropical and subtropical parts of the world in the late Miocene, measured directly from the isotopic composition of the tooth enamel of grazing herbivores. Continuing into the Pliocene of Africa, this trend led to an increase in the proportion of herbivores of savanna-mosaic aspect. The ancestral form of the white rhinoceros is known from late Miocene deposits, perhaps as a new immigrant from Arabia, and giraffes made their first appearance, as obvious immigrants from Asia. From around 6.5 Ma, we see the rise to prominence of the hippotragine, reduncine, and tragelaphine antelopes,

FIGURE 5.5

A selection of carnivores from the late Miocene of eastern Africa

(*From left to right*) The hyena *Ictitherium ebu*, the amphicyonid bear-dog *Amphicyon*, the machairodont cat *Dinofelis*, the mustelid *Ekorus ekakeran*, the viverrid *Viverra leakeyi*, and the machairodont cat *Lokotunjailurus emageritus*.

Each square measures 1 m on a side.

together with impalas, and perhaps somewhat later the alcelaphines. It is during this period, too, that the more dog-like hyenas, the genera *Ictitherium*, *Hyaenictitherium*, *Lycyaena*, and *Hyaenictis*, appear, together with the felid genera *Dinofelis* and *Machairodus*, large and very large cats. In all, a significant further immigration from Eurasia is indicated, even if *Dinofelis* may have had its origins in Africa. Among the mustelids, the genus *Ekorus* from Kenya was big enough to deserve inclusion in the guild of larger carnivores (figure 5.5).

Plio-Pleistocene: Faunal Turnover and the Appearance of the Modern Fauna

Into the Pliocene, the effects of continued uplift operated in combination with further aridification as the southern ice sheet grew on Antarctica, the global sea level dropped, and the Mediterranean became a landlocked sea that gradually evaporated. The Mediterranean eventually regained contact with the Atlantic Ocean, but the process of change in Africa was continued by the swing to even colder global temperatures after 3 Ma. The Sahara became a desert, and by 2.5 to 2 Ma, the effect on the African biota was marked. The paleontologist Elisabeth Vrba has identified this late Pliocene event as a "turnover pulse," a change in physical environment that induces significant evolutionary transformations across numerous lineages. These evolutionary changes involved extinctions, speciations, dispersals, and modifications within lineages, altering both the composition of the larger fauna and the morphology of the species—exemplified by the relationship between changes in ungulates and those in predators.

Ungulates

In discussing the African ungulates, we may usefully start with the elephants. Throughout the Pliocene of eastern Africa, they showed a trend toward extremely high-crowned and folded teeth with extensive enamel cutting ridges, with the rate of this change increasing markedly in the later Pliocene, especially after around 2.3 Ma. At that time, the lineage represented by the extinct *Elephas recki* became the dominant species, while the forerunners of the living African elephant genus, *Loxodonta*, declined. This pattern of massive dental development also is seen in the pigs, several species of which underwent increases in the size of the teeth and in the height of the crown during the Pliocene, with the rate of change becoming more rapid at this time. There also was considerable turnover in the suid fauna of the later Pliocene in Africa between around 2 and 1.6 Ma—what the paleontologist Tim White has referred to as "the most dramatic faunal turnover" of suid evolution—with the extinction of members of the *Nyanzachoerus–Notochoerus* lineage, *Notochoerus euilus* and *N. scotti*; the appearance of *Kolpochoerus majus*; the disappearance of *Metridiochoerus andrewsi*; and the emergence of *M. modestus*, *M. hopwoodi*, and *M. compactus* (figures 5.6 and 5.7).

The pigs were an extremely important part of the overall diversity of the Plio-Pleistocene fauna of Africa, before being reduced to the three extant taxa, and are likely to have been equally important to the carnivores as prey. Today, warthogs may figure largely in the diet of lions, especially in areas where other prey are scarce or temporarily absent. However, some of the extinct species, especially those of the genus *Notochoerus*, were huge animals that weighed up to perhaps 450 kg and therefore were relatively immune to predation, although the young may have been a valuable source of food for an open-country predator, such as *Homotherium*. Large piglets of larger species would have represented a good return on energy expended in capture, and a turnover in the suid component of the ungulate fauna that removed such larger pigs from the species mix would have had important consequences for predators.

Horses of the living genus *Equus* first appeared in Africa around 2.3 Ma, following their dispersion across Eurasia from the Americas via the Bering Land Bridge, and the high-crowned teeth of these animals were well equipped for the consumption of abrasive fodder. The indigenous equids of the extinct hipparionine group coexisted

FIGURE 5.6

A selection of ungulates from the Pliocene of eastern Africa

(*From left to right*) The hipparionine *Eurygnathohippus cornelianus*, the giraffe *Giraffa jumae*, the trage-laphine antelope *Tragelaphus nakuae*, the pig *Kolpochoerus limnetes*, the giraffid *Sivatherium maurisium*, the impala *Aepyceros melampus*, the chalicothere *Ancylotherium hennigi*, the hippopotamus *Hexaprotodon aethiopicus*, and the rhinoceros *Ceratotherium praecox*.
Each square measures 1 m on a side.

with the newcomers for some time and showed their own interesting pattern of dental development, matching that of the elephants and pigs during the Pliocene. At the same time, the white rhinoceros (*Ceratotherium simum*) exhibited changes in the structure of the skull, which became longer and enabled the species to graze on shorter grasses more easily, features duplicated to some extent among hipparionine horses of the genus *Eurygnathohippus*.

Antelopes, the most taxonomically diverse of the larger mammals of the African Plio-Pleistocene, have long been an important component of the fauna of the continent and are likely to have been among the most significant prey species sought by all the larger predators. Elisabeth Vrba has identified a major change in the antelope fauna between 3 and 2 Ma, involving speciation, immigration of antilopine and caprine taxa, and extinction. The event appears to have been a two-stage, "relay"

FIGURE 5.7

A selection of ungulates from the Pleistocene of eastern Africa

(*From left to right*) The rhinoceros *Ceratotherium simum*, the giraffe *Giraffa camelopardalis*, the zebra *Equus koobiforensis*, the bovine *Pelorovis olduwayensis*, the alcelaphine antelope *Megalotragus issaci*, the antilopine antelope *Antidorcas recki*, the reduncine antelope *Menelikia lyrocera*, the pig *Metridiochoerus andrewsi*, the hippopotamus *Hippopotamus gorgops*, and the giraffid *Sivatherium maurisium*. Each square measures 1 m on a side.

turnover pulse. As the climate cooled, first the warmth-, woodland-, and moisture-preferring taxa changed, with the appearance at around 2.7 Ma of more species in an effort to keep up with shifting conditions. By 2.5 Ma, the new taxa were those with adaptations to more open terrain, including an increase in the importance of the open-country, *larger* bovids, especially among the alcelaphines and hippotragines. They may have been somewhat faster, or at least more efficient in their locomotion, adaptations that may be linked to the extension of the foraging range or perhaps to the need to avoid predators in more open country. These changes in the structure of the antelope fauna itself were matched by changes in the dentition in later species of lineages, such as a shift in emphasis to molars at the expense of premolars in alcelaphines like *Connochaetes*, the wildebeest. Such modifications often involve the loss of the lower second premolar and the reduction of the third premolar to a peg-like structure, an emphasis on the molar tooth row that seems to indicate increasing adaptation to coping with the food available in more open grasslands.

Analyses of postcranial characters suggest that some of the locomotor changes took place slightly later, but investigations of bovid feeding mechanisms tend to support arguments for environmentally induced turnover in the period before 2 Ma. Further changes occurred in the bovids after 2 Ma, with immigrations of yet more open-country species of ovibovines and caprines seen in various horizons in the Omo Group: at West Turkana and in Member H of the Shungura Formation. Such changes in the structure of the bovid guild, like those among the pigs, would have directly affected members of the large-carnivore guild.

Carnivores

When we examine the record of the true carnivores, the members of the order Carnivora, we can distinguish a number of distinct phases in the evolution of the African fauna, particularly over the past 5 Ma. The large-predator guild has changed markedly over that period; in the early Pliocene, there were no living species of cats, dogs, or hyenas, and the hunting and scavenging roles were occupied by saber-toothed cats, medium-size hunting hyenas, and smaller hyenas with some moderately well developed bone-crushing abilities.

By around 3.5 Ma, many of the extant species of large carnivores were present in a rich guild almost twice the size of that of today. For the next 2 Ma, modern elements, with one extreme of hunting adaptation in the form of the cheetah among them, coexisted with some of the more archaic carnivores, such as the machairodont cats of the genera *Homotherium*, *Megantereon*, and *Dinofelis*, as well as the so-called hunting hyenas of the genus *Chasmaporthetes* and the giant hyena *Pachycrocuta* (figure 5.8). The extinction of the saber-toothed cats left the living cats as the sole felid members of the guild—a particularly important change.

FIGURE 5.8

A selection of carnivores from the Pliocene of eastern Africa

(*From left to right*) Lion (*Panthera leo*), cheetah (*Acinonyx jubatus*), leopard (*P. pardus*), spotted hyena (*Crocuta crocuta*), the machairodont cat *Dinofelis*, striped hyena (*Hyaena hyaena*), and the machairodont cats *Megantereon cultridens* and *Homotherium latidens*.
Each square measures 1 m on a side.

The larger modern cats capture, kill, and then eat in a three-stage process in which a killing bite in the neck region to induce blood loss and shock, or a clamping of the muzzle, causes suffocation during capture and despite any struggles of the prey. This is as true of cheetahs as it is of lions and leopards. Several interrelated features of anatomy must have both constrained and directed the hunting strategy and killing tactics of machairodonts in another direction. Both *Dinofelis* and *Megantereon* had a strong body with a short back, large claws, and—especially in *Megantereon* and *Dinofelis piveteaui*—long, sharp but fragile canine teeth. These cats were built for capturing prey by means of a short rush and considerable strength to bring down and hold an animal before killing it by some form of slashing bite. Any attempt to clamp the muzzle would have risked biting into bone, and only with prey held essentially still could the machairodonts have slashed or bitten into the neck without possibly damaging the canines as a result of striking bone or causing torsion.

Homotherium also had elongated, flattened, and therefore potentially breakable upper canines as well as a strong back, although its body pattern differed from that of the other machairodonts in having more slender limbs. The front limb, in particular, was more elongated and the claws appear to have been relatively small, the body plan of a cursorial hunter and perhaps indicative of group-hunting behavior. Hunting by chasing can be done only in open terrain, where larger group size may be necessary to deter scavengers. Such group activity is clearly indicated in the American species *Homotherium serum*, known in some numbers from the late Pleistocene site of Friesenhahn Cave in Texas in association with numerous remains of juvenile mammoths. In short, none of these cats could have killed in the same way as a lion or a leopard, with the attendant risk of damage to its teeth, and the most probable hunting habitat for *Megantereon* and *Dinofelis* was relatively closed vegetation, permitting a close stalk of a carefully selected prey animal followed by a short dash from cover and firm seizure. For *Homotherium*, group hunting in more open areas of vegetation would have achieved the same end, with the prey subdued by weight of numbers.

The machairodont cats, together with the hyenas of the genera *Pachycrocuta* and *Chasmaporthetes*, became extinct in Africa around 1.5 Ma, although regional variations undoubtedly occurred. But what was the nature of the interactions between them and the living predators, particularly between the extinct and the extant cats, during the period when they overlapped? If we look at the relationships between species of modern large cats and their interactions with other carnivores, we can get several useful clues. One interesting point to consider is the ecological separation of, and hierarchical relationship between, lion-size and leopard-size cats. In areas where the tiger and the leopard are able to coexist, it seems that vegetational cover is a key factor. Prey species in the weight range of 25 to 50 kg are relatively abundant and can be consumed quickly by the leopard before competitors arrive or be easily transported and hidden or carried into a tree. Where cover is sparse, the leopard finds it increasingly difficult to avoid encounters with the bigger tiger, which steals carcasses and even kills the small leopard, so it becomes scarcer or simply disappears. In areas where the tiger is now absent, such as Sri Lanka, the leopard ventures into open terrain in pursuit of larger prey, even in broad daylight. In Africa, the leopard also seeks refuge in more wooded areas, particularly gallery forests, but also can thrive in relatively dry, open environments, such as the Kalahari Desert. The factors that influence hunting success, including the extent of cover, appear to be the same for both lions and leopards, but the lions can survive in sparsely covered areas by means of group action.

It appears likely that extinct cats of lion size, such as *Homotherium*, would have been similarly dominant over smaller species, such as *Dinofelis* and *Megantereon*, and the presence of the larger taxon associated with one or both of the others at a fossil locality would suggest the presence of enough cover in a mosaic of vegetation for the smaller species to avoid encounters with the larger. But the complex and shifting nature of the relationship between habitat and predator activity must be borne in mind. Even modern predators well adapted for life on the plains will suffer a decline in population with a decrease in dry-season rainfall, as ungulates migrate and competition increases for the remaining resident prey. This competition appears to operate largely within species, but has particular implications for interspecies rivalries.

Evolutionary events in the African predator and prey components of the mammalian fauna happened in the aftermath of the changes in vegetation changes around 1.7 Ma, part of a global phenomenon marked by shifts in the deep-sea-sediment core record of oxygen isotope values at 1.9 Ma and perhaps ultimately those at 2.5 Ma. Earth as a whole had become colder and probably drier in a long-term trend marked by steps. The severity of the climatic event is underlined by the extinction in Africa of three long-lived machairodont cat genera and perhaps three hyena species in its aftermath. *Megantereon* and *Dinofelis* would have been badly affected by reductions in vegetational cover, changes in size and antipredator response of prey species (especially among the antelopes and pigs), and perhaps increasingly frequent encounters in more open terrain with such large competitors as *Homotherium*, lions, and hyena clans. *Homotherium* is likely to have been badly affected by reductions in resident prey biomass in successive years of increased dry-season drought and by increasingly frequent clashes with open-terrain predators. But we should note that it took the combined effects of changes in climate, vegetation, and prey composition and structure to drive the machairodonts to extinction in Africa, and even then they managed to continue to exist on other continents—in the Americas to within the past 20,000 years. Theirs was no evolutionary "flash in the pan," no short-term success story for a subfamily of what have been characterized as bizarrely adapted animals. They are extinct, but so are most organisms that have ever lived.

The Modern Mammalian Fauna

The turnovers of the African mammalian lineages that occurred toward the end of the Pliocene and into the earliest part of the Pleistocene may be regarded as the final steps in the appearance of the living fauna of the continent. Of course, changes in the composition of faunas have been global phenomena, and, as mentioned, virtually none of the living species of mammal had yet appeared by 5 Ma—a measure of the level of evolution that has occurred in a relatively short period. As well as the alterations that occurred over that time span, Europe, Asia, and the Americas witnessed massive changes in their large-mammal faunas as recently as the end of the last ice age, between 20,000 and 10,000 years ago, when impressive numbers of animals became totally or locally extinct. North American examples include mammoths, mastodons, camels, species of deer and bears, horses, saber-toothed cats, a cheetah-like cat, and lions; Eurasia lost mammoths, woolly rhinoceros, musk oxen, species of deer and bears, hyenas, lions, and leopards. The modern mammalian fauna of these areas, in other words, is a very recent, postglacial group with a much reduced number of species.

This is not true of Africa. Although a relatively small number of species—mainly ungulates—became completely or locally extinct at the very end of the Pleistocene, the changes that occurred during the earliest part of that epoch were virtually the last of any major significance. Once the saber-toothed cats disappeared around 1.5 Ma, the large-predator guild had taken on its present structure and composition (figure 5.9), and this modern stamp began to be reflected in the fauna as a whole. The archaic-looking deinotheres, gomphotheres, and chalicotheres disappeared, as did the three-toed hipparionine horses, while the pigs and giraffes were eventually reduced in number and diversity to the present small range of species. Among our own fossil relations, as we will see, the robust australopithecine hominins of the genus *Paranthropus* were among the larger primates that went extinct. At the same time, however, species that had appeared on the continent during the later Pliocene have survived in greater numbers down to the present day. Africa, therefore, began to look modern at a substantially earlier date than Eurasia or the Americas, while still retaining a larger proportion of its earlier range of larger mammals.

The extinctions that occurred in Africa at the end of the Pleistocene involved no more than perhaps 10 complete disappearances of species, such as the long-horned buffalo (*Pelorovis antiquus*), which had been widely recorded in northern, eastern, and southern parts of the continent, and the giant hartebeest (*Megalotragus priscus*).

FIGURE 5.9

A selection of carnivores from the Pleistocene of eastern Africa

After the extinction of the machairodont cats and the more archaic hyenas, Africa was left with the modern guild of larger carnivores: (*from left to right*) cheetah (*Acinonyx jubatus*), leopard (*Panthera pardus*), lion (*P. leo*), spotted hyena (*Crocuta crocuta*), striped hyena (*Hyaena hyaena*), black-backed jackal (*Canis mesomelas*), and African hunting dog (*Lycaon pictus*).
Each square measures 1 m on a side.

Localized extinction was a far more common phenomenon, however, with animals like elephants, zebras, and giraffes disappearing from the Maghreb; kobs and lechwes from Sudan; and white rhinos, black wildebeest, and springboks from the Cape region. Human activity has been invoked to explain some of or all these late extinctions, in part because such species clearly survived earlier fluctuations in environment and habitat of the kind that occurred at the end of the last ice age, but the evidence is circumstantial rather than direct.

Human Evolution

Human evolution is clearly of particular interest to us. But we must reiterate that the evolution of humans in Africa took place as an integral part of all other aspects of the evolution of the primates and of the biota in general, although it does present its own suite of difficulties for analysis. A map of the most important localities with hominin remains is shown in figure 5.10, while a timescale for the appearance of the various taxa is given in figure 5.11.

Earliest Stages

The early record of human evolution, as we briefly discussed, presents a complex series of morphologies and proposed taxa, and interpretations of adaptations, iden-

FIGURE 5.10

The distribution of hominin sites in Africa

The hominin sites in Africa: *1*, Lothagam, Kenya; *2*, Aramis, Ethiopia; *3*, Kanapoi, Kenya; *4*, Chad Basin: Toros-Menalla and Koro Toro; *5*, Laetoli, Tanzania; *6–8*, Turkana Basin: Omo River, Ethiopia, and East Turkana and West Turkana, Kenya; *9*, Chiwondo Beds, Malawi; *10*, Hadar, Ethiopia; *11*, Bouri, Ethiopia; *12*, Bodo D'Ar, Ethiopia; *13*, Olduvai Gorge, Tanzania; *14*, Taung, South Africa; *15*, Sterkfontein Valley: Sterkfontein, Swartkrans, and Kromdraai, South Africa; *16*, Drimolen Cave, South Africa; *17*, Gondolin, South Africa; *18*, Makapansgat Limeworks, South Africa; *19*, Olorgesailie, Kenya; *20*, Ternefine, Algeria; *21*, Kabwe or Broken Hill, Zambia; *22*, Klasies River Mouth, South Africa.

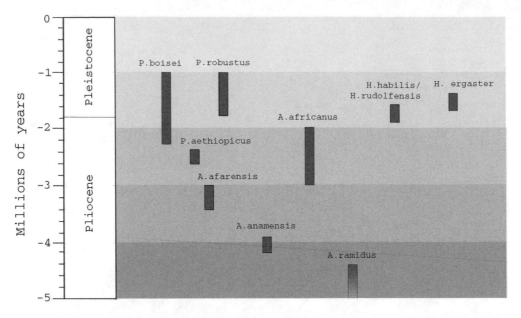

FIGURE 5.11

Timescale of the appearance of early hominin species

This very schematic diagram provides a quick visual overview of the time span of some of the more established African hominin species of the later Pliocene and early Pleistocene. We have deliberately avoided any effort to show links between the various taxa, although the likely separation of the "robust" australopithecines of the genus *Paranthropus* is clear.

tity, and relationship of the various species are made no easier by the fragmentary state of much of the material and the frankly combative nature of some of the discussion. In very broad terms, our African ancestors and relatives changed from being generalized apes to being more sophisticated apes with tools over a few million years, between perhaps 7 and 2.5 Ma. They became really recognizably human only around 1.8 Ma, with the earliest representative of the *Homo erectus* lineage. This transition included shifts to an upright stance and fully bipedal walking, evidenced in both morphology and the footprint evidence from Laetoli in Tanzania, as well as a massive increase in relative and absolute brain size. We must presume that these physical manifestations of evolution were accompanied by alterations in behavior, social interactions, and intelligence. What we do not know, however, is the extent to which the changes in nonphysical attributes paralleled those in physical characteristics, such as locomotion and brain size.

Traditional interpretations of this larger evolutionary pattern of ape-like to human-like primate have seen it as a reflection of an adaptation to an increasingly terrestrial lifestyle that included the eating of meat, and the archaeological record appears to

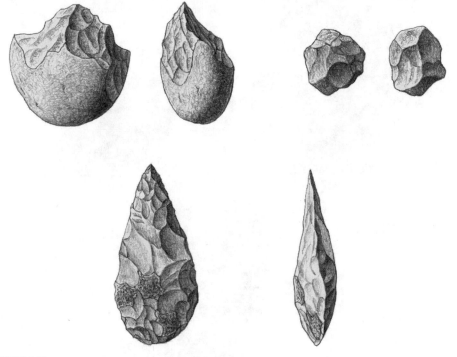

FIGURE 5.12

Some early stone tools

(*Top left*) Two views of an Oldowan chopper. (*Top right*) Two views of an Oldowan polyhedron. (*Bottom*) Two views of an Acheulean hand ax. While the Oldowan tools are crude, seemingly expediently made and simply functional, the Acheulean implements often are very beautifully made, seemingly far beyond the point of sheer necessity. However, the process of making hand axes produced useful, sharp flakes, and some anthropologists have suggested that an ax could have functioned as a portable source of easily detached tools of this kind rather than as a finished product.

support such analyses. According to interpretations of the material referred to *Ardipithecus ramidus*, the attachment to the woodland habitat may have persisted until almost 4 Ma. The development of a stone-tool technology can now be traced back in Africa to around 2.4 Ma at Hadar and in the Shungura Formation in Ethiopia and in the Lokalalei Member at West Turkana in Kenya, while the bones of large mammals from Bouri in Ethiopia have damage marks interpreted as having been made by stone tools. What happened in the intervening period either to assist or to motivate the move to more open terrain is unclear, however, although it is always possible to suggest some increasing use of a technology involving unpreserved organic materials, such as wood.

The earliest stone tools, often referred to as Oldowan, based on their initial dis-

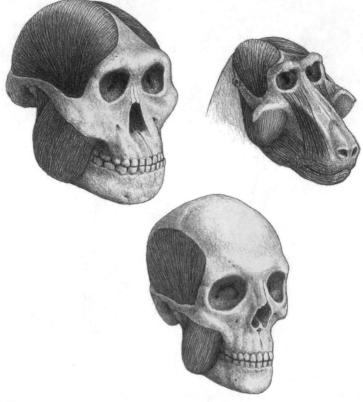

Comparison of the skulls of *Paranthropus robustus*, *Theropithecus brumpti*, and *Homo sapiens*

Some adaptations of the skull, such as the anterior projection of the insertion area for the masseter muscles in the malar bone, evolved in parallel in robust australopithecines, such as *Paranthropus robustus* (*top left*), and baboons of the gelada group, like *Theropithecus brumpti* (*top right*). These adaptations, coupled with well-developed cheek teeth, indicate that both primates had to exert great crushing force not only at their most posterior molars, but also at their more anterior premolars. Compared with those of both fossil species, the skull of a modern human (*Homo sapiens*) (*bottom*) not only lacks the anterior expansion of the malars, but also has much weaker masticatory muscles.

covery at Olduvai Gorge in Tanzania, were essentially sharp-edged flakes and perhaps pounding implements (figure 5.12), primitive in form but undoubtedly useful to a primate with the dexterity to employ them and certainly suitable for processing a range of food items, including the carcass of an ungulate. Such tools were augmented by around 1.4 Ma by more sophisticated, bifacially worked implements generally referred to as Acheulean and commonly found in the form of what are known as hand axes.

If both *Homo habilis* and *H. rudolfensis* are now to be removed from the genus *Homo*, then logically the earliest stone tools were made by members of another genus, unless we propose the existence of one or another still unknown species of our own

FIGURE 5.14

Size comparison of *Theropithecus oswaldi*, *Paranthropus boisei*, and *Homo erectus*

The accumulation of fossils of the giant gelada *Theropithecus oswaldi* at the site of Olorgesailie in Kenya has been regarded as evidence that *Homo erectus* (*ergaster*) was hunting these baboons in the Pleistocene. If that were the case, it would have been a remarkable feat for the hominins, considering that an adult male *T. oswaldi* (*left*) could be nearly 80 cm tall at the shoulder and as heavy as a female gorilla. As this size comparison shows, just being faced by such an impressive monkey would have been testing enough for even a large hominin like *H. erectus* (*right*). For a smaller hominid, like the contemporary *Paranthropus boisei* (*center*), the giant baboon would have been a formidable opponent, possibly competing to some degree for similar resources.

genus. But toolmaking by hominins outside the lineage of *Homo* presents no difficulties for our understanding. It is widely recognized that chimpanzees can and do both make and use tools, however primitive, sufficient to satisfy any sensible definition of such behavior. Thus there is nothing strange in according tool-using status to members of another evolved primate genus. Since we are unclear about the nature of the precise relationship between earliest *Homo* and the australopithecines *Paranthropus* and *Australopithecus*, we are free to suggest that those hominins who made tools were closest to ourselves and perhaps more directly ancestral to the *Homo* lineage, but, of course, that supposition is not necessarily true.

What is clear about the fossil and archaeological evidence is that it cannot be interpreted to show a simple, linear sequence of development in either hominin anatomy or hominin technology that takes us from ape-like to human-like by way of the various forms seen in the fossils. Indeed, it becomes evident that a variety of evolutionary solutions were developing among the African hominins during the Pliocene, and this variety may well extend back into the Miocene if the material referred to *Orrorin tugenensis* and *Sahelanthropus tchadensis* bears continued scrutiny. The course of our own lineage seems to have involved an evolution toward greater ecological generalization coupled with an emphasis on stone-tool technology and an increase in brain size, as Elisabeth Vrba pointed out several years ago. At the same time, the robust australopithecines of the genus *Paranthropus* appear to have followed a trend toward larger jaws and teeth, presumably coping with food acquisition and processing by means of such an anatomical development, what is referred to as hypermasticatory ability (figures 5.13 and 5.14). At least one hominin species traditionally placed in the genus *Homo*, the East African *Homo rudolfensis*, appears to have followed more or less the same path as the paranthropines between 2.5 and 1.8 Ma, with its own enlarged teeth; even if it is removed from *Homo*, we cannot escape the fact that it does have a relatively large brain as well. If the *H. rudolfensis* material is indeed to be linked with the earlier *Kenyanthropus platyops*, we may very well have evidence of a separate lineage within which both brains and teeth developed. Since we have no direct evidence to link any particular Pliocene hominin with the most primitive of stone tools, it may well be that such a technology was employed by several of the later Pliocene species.

We may therefore identify at least three evolutionary developments in Pliocene hominins, but the most interesting point about them is that all may be regarded as responses to one problem: dealing with the changing climate and habitats faced not only by the hominins, but by all the members of the mammalian fauna. The only lineage to survive ultimately was that of our own genus, *Homo* (figure 5.15), but the various species of *Paranthropus* had a very reasonable measure of success, too, before going extinct around 1 Ma in both southern and eastern Africa. Perhaps they managed some form of toolmaking to augment their physical capabilities (figure 5.16). Of course, all the species of hominin other than *Homo sapiens* are now extinct, but with

---

FIGURE 5.15

Three vignettes of a group of *Homo erectus*

Do the origins of language lie in the use of an increasingly complex technology and the social behavior that accompanies it? Our ancestors may have honed their communication skills as they worked cooperatively to make stone tools (*top*) and use them to butcher an antelope carcass, while one group member stands guard (*center*), and, finally, relaxed in the shade of a tree.

FIGURE 5.16

Two paranthropines attempt to make tools

The known anatomy of the hand of "robust" australopithecines, or paranthropines, indicates that they were as well adapted physically for tool making as members of our own genus, *Homo*. A different matter is whether they ever did make tools, either from invention or from imitation of actions observed in other hominins. With several species of the genus *Paranthropus* and members of the genus *Homo* living sympatrically in eastern and southern Africa during the Plio-Pleistocene, it is very difficult to be sure about the precise makers of stone tools.

continued debate over the pattern of relationships we are unsure where any particular lines of genetic continuity may lie.

Dispersions

The earliest known hominin species appear to have been quite restricted in their distributions in Africa, with *Ardipithecus ramidus*, *Australopithecus afarensis*, *A. africanus*, and species of the robust genus *Paranthropus* being variously recorded only from eastern and southern parts of the continent. If arguments for the association between earliest hominins, such as *Ardipithecus*, and more wooded environments are correct, the distribution of such habitats would, of course, have been a major determinant of the range of the hominins. The specimens traditionally referred to *Homo rudolfensis*, and now perhaps to *Kenyanthropus*, are also limited to eastern Africa, although the sample of material assigned to *H. habilis* may include remains from the southern region, and these distributions may tell us something about the centers of origin and the subsequent dispersal patterns of the various taxa. Of course, though, it is easy to read too much into the face-value distribution of what are, after all, relatively scarce fossils, especially when each newly discovered one tends to be claimed as a new species or genus that completely changes our view of human evolution. But there are good evolutionary reasons for expecting sexually reproducing species to originate in a restricted area, one with a preferred habitat to which the shared system of mate recognition is closely adapted. Any greater range for a species must then be achieved by dispersion, and the species with the most general repertoire of adaptations may be in the best position to disperse most rapidly and most successfully. It may be argued that the development of a stone-tool technology was one of the ways in which some of the hominins, including members of our own lineage, began to extend the range of their adaptive skills toward the end of the Pliocene. In the absence of such technology, earlier hominin species would have been more constrained in their options for dispersion.

The question of the timing of the earliest emergence of humans from Africa remains the subject of intense debate. Claims and counterclaims for an early appearance in this or that country have been made over the years, but the best attested and most widely accepted evidence still is that from 'Ubeidiya in Israel, dated to around 1.4 Ma, and from Dmanisi in Georgia, dated to between 1.8 and 1.6 Ma. The human remains from 'Ubeidiya are meager, but informative, and are reinforced by the presence of Acheulean stone tools. Dmanisi has provided a fine lower jaw with teeth and, more recently, a pair of excellent though damaged crania and a smaller, almost complete skull with mandible, but the stone-tool assemblage from this site has no Acheulean affinities. Although the mandible is difficult to assign to any particular species with certainty, the two crania are very reminiscent of specimens from Turkana Basin localities of similar age, especially the juvenile *Homo erectus* from West Turkana. The almost complete skull, however, is of a smaller individual, raising the question of whether it simply represents a female of a fairly dimorphic species or an individual of a species different from that represented by the two crania. Such uncertainty bedevils paleontologists, but the most plausible interpretation surely has to be that the fossils are those of a single species. The most likely candidate at both these early Eurasian localities, then, has to be *H. ergaster* or *H. erectus* (figure 5.17). Nevertheless, such finds say relatively little about the true earliest date of human emigration from Africa. For one thing, they are unlikely to record the earliest movement and almost certainly do not do so. To proceed further in understanding the timetable of the dispersion of humans, we must look at the wider pattern of the migration of mammals between Africa and Eurasia.

The Arabian Peninsula, an arid region closed off to the north by the great arcs of the Taurus and Zagros Mountains of present-day Turkey and Iran, would have presented a formidable obstacle to interchanges. The evidence for migration by non-

FIGURE 5.17

A group of *Homo erectus* looks beyond Africa

At some point in the early Pleistocene, a group of early humans must have been the first to reach north-eastern Egypt and look beyond Africa to the wider world.

human mammalian species suggests that this barrier was at its most impenetrable during the earliest Pleistocene. In contrast, the later part of the Pliocene appears to have been a time of relatively easy movement across the region, with various Eurasian antelope species and the horses of the genus *Equus* successfully colonizing Africa, and mammoths and some African antelopes moving out to the north. If hominins were also dispersing in the late Pliocene, the earliest ones to leave Africa may not even have been members of the genus *Homo*.

Whatever the precise timetable of the earliest emergence of humans from Africa, it is now clear that the worldwide appearance of fully modern peoples—that is, *Homo sapiens*—was a much later phenomenon. The first emigrants from Africa eventually reached most parts of Eurasia, but all indications are that they were not the direct ancestors of modern humans. It is now generally accepted that the late appearance of our species, by around 150,000 years ago, also occurred in Africa, represented by the group of fossils from Border Cave and Klasies River Mouth in South Africa and from Omo-Kibish in Ethiopia.

EPILOGUE

WE SET OUT TO FOLLOW the evolution of the African mammalian fauna over
some 35 Ma against the background of changes in the topography, vegetation, and climate of the continent—in short, the evolution of the African Eden. Today, Africa is
one of the richest biotic regions on Earth, a zoogeographic realm in itself, with
around one-quarter of all the living species of mammals and a vast array of plants.
Yet we can now see that this abundance of animals shrinks in comparison with the
range of species that have come and gone on the continent over a geologically short
period. For all its rich mammalian diversity—with its lions, cheetahs, elephants,
antelopes, giraffes, rhinos, and zebras—an Africa without saber-toothed cats or giant
hyenas; without deinotheres, three-toed horses, or chalicotheres; without pigs the size
of a small cow or sivatheres; or even without robust australopithecines—not just
instead of the extant fauna but, in most cases, *in addition to it* within the past few million years—is a much diminished place. The eighteenth-century Scottish geologist
James Hutton argued that the present is the key to the past, that we need have no
recourse to miracles or other divine interventions to explain the evolution of the
world. True enough, but even a glance at the fossil record shows that while the present may be the key to the past in terms of mechanism and process, in many ways it
is also but a pale shadow of it in terms of richness and diversity.

What will be the fate of the survivors? It has become fashionable to argue that bio-
diversity in general is worth preserving because it may be of direct benefit to humans,
particularly in the field of pharmacology. Species as yet undiscovered may prove to
be the solution to this or that problem, perhaps to provide the miracle cure for an ail-

{ 241

ment that besets us. That may be true and, if so, certainly provides a good reason for us to exercise restraint in our treatment of the African biota. The sad fact is that a direct return—in terms of money, immediate benefit in kind, or votes—is an increasingly important guiding principle of business and politics alike. But there is another, more fundamental strand to the argument: while the biota of Africa has changed and its diversity has shrunk, the continent still is home to the longest-enduring mammalian fauna, with a higher proportion of Pliocene species than remains on any other continent. This richness is of long-standing because it is the product of a biomic patterning whose origins can be traced back to the Miocene. Put most simply, the modern mammalian fauna of Africa is the oldest in the world. One of our fossil relatives from among the australopithecines visiting the Serengeti today might well comment on the absence of this or that species, but would still happily recognize virtually all the animals on the plain. We cannot be so careless as to let that situation change— can we?

Science offers little or nothing in the way of moral certainties, but it does provide data on which to base ideas and reach decisions. Of course, we might choose to look at the fossil record and conclude that extinction has been a general feature of life on Earth and that we accept no responsibility for it through our past, present, or future actions; while we may be a product of the evolution of the African fauna, we are no more guardians of it than is any other species. But we surely cannot escape the outcome of our own evolution so easily. Our developed abilities, not least those that enable us to recover the past as well as alter the future through technology, now put us far beyond being simply one more member of the fauna, beyond being a species able to record and comment but content to stand back. While relatively few of the complete disappearances in Africa during the latest Pleistocene can be clearly blamed on humans, it is evident that our interference has played a major part in altering the distributions of numerous animal species and has brought a significant number to the verge of extinction. As the twentieth century alone has shown us, there is really very little doubt that human development has the potential to destroy, and to do so on a massive scale.

Biologists and paleontologists, of course, and, for that matter, tourists, may have a perspective on the diversity of the African mammalian fauna, and the desirability of maintaining it, very different from that of the people who live in the region. Any efforts to preserve biotic diversity in Africa must take account of and involve those who live there, people whose crops, animals, and very lives are often directly affected by the depredations of leopards, lions, baboons, pigs, or elephants. All these animals can wreak havoc in a field of crops or among a herd of domestic animals, and a disenchanted or simply hungry local population of farmers or ranchers will hinder progress for any conservation program. Game reserves to which animals are restricted, and from which humans are excluded except as tourists or scientists, is one proposed solution, but others argue for education and a share in the profits of managing wildlife on a larger scale outside such reserve areas. In his book *The African Leopard*, the biologist Theodore Bailey quotes a 1988 figure of $29 million as an estimated exploitation value for the cat alone to traders and so-called sport hunters. While employing trophy hunting as a means to ensure the survival of a species may strike many as distasteful, there is undeniably a market for reducing animals to skin and horns. But such an approach to conservation would surely imply that only the most beautiful or powerful animals would be preserved simply to be shot, for who among the sporting fraternity would pay to bag a spotted hyena?

Whatever decisions we as a species make about the future of African wildlife, we should recognize that time is passing while we continue to weigh the arguments and achieve some perfect justification and solution. In his excellent book *The Variety of Life*, the naturalist and writer Colin Tudge argues that all such attempts to justify conservation based on economics must fail, as must those based on morality, since the logical underpinnings of ethical values tend to be equally illusory. The only basis for action, in his view, is emotional response to the problem, because strong feelings will find practical reasons for their support. If there is any truth in that argument, we may hope that some of what we have presented in this book may help engender such an emotional response. If people will not pay to shoot a spotted hyena or another equally unglamorous animal and therefore, by that logic, help preserve the species, perhaps they will instead appreciate the extent to which hyenas have been among the major players in the evolution of the African mammalian fauna and go on to argue for their place in its future. If we preserve hyenas, there may be hope for the rest.

Further Reading

CLIMATE AND CLIMATE CHANGE

Cane, M. A., and P. Molnar. 2001. Closing of the Indonesian seaway as a precursor to East African aridification around 3–4 million years ago. *Nature* 411:157–162.

de Menocal, P. B. 1995. Plio-Pleistocene African climate. *Science* 270:53–59.

de Menocal, P. B., and J. Bloemendal. 1995. Plio-Pleistocene climatic variability in subtropical Africa and the paleoenvironment of hominid evolution: A combined data model. In E. S. Vrba, G. H. Denton, T. C. Partridge, and L. H. Burckle, eds., *Paleoclimate and Evolution, with Emphasis on Hominid Origins*, pp. 262–288. New Haven, Conn.: Yale University Press.

Denton, G. H. 1999. Cenozoic climate change. In T. G. Bromage and F. Schrenk, eds., *African Biogeography, Climate Change, and Human Evolution*, pp. 94–114. New York: Oxford University Press.

MacLeod, N. 1999. Oligocene and Miocene palaeoceanography—A review. In P. J. Whybrow and A. Hill, eds., *Fossil Vertebrates of Arabia*, pp. 501–507. New Haven, Conn.: Yale University Press.

Partridge, T. C., G. C. Bond, J. H. Hartnady, P. B. deMenocal, and W. F. Ruddiman. 1995. Climatic effects of late Neogene tectonism and volcanism. In E. S. Vrba, G. H. Denton, T. C. Partridge, and L. H. Burckle, eds., *Paleoclimate and Evolution, with Emphasis on Hominid Origins*, pp. 8–23. New Haven, Conn.: Yale University Press.

Prentice, M. L., and G. H. Denton. 1988. The deep-sea oxygen isotope record, the global ice sheet system and hominid evolution. In F. E. Grine, ed., *The Evolutionary History of the "Robust" Australopithecines*, pp. 383–403. New York: Aldine de Gruyter.

Shackleton, N. J. 1995. New data on the evolution of Pliocene climatic variability. In E. S. Vrba, G. H. Denton, T. C. Partridge, and L. H. Burckle, eds., *Paleoclimate and Evolution, with Emphasis on Hominid Origins*, pp. 242–248. New Haven, Conn.: Yale University Press.

Stanley, S. M. 1998. *Children of the Ice Age*. New York: Freeman.

EVOLUTION AND EVOLUTIONARY MOTORS

Grubb, P. 1999. Evolutionary processes implicit in distribution patterns of modern African mammals. In T. G. Bromage and F. Schrenk, eds., *African Biogeography, Climate Change, and Human Evolution*, pp. 150–164. New York: Oxford University Press.

Janis, C. M. 1993. Tertiary mammal evolution in the context of changing climates, vegetation, and tectonic events. *Annual Review of Ecology and Systematics* 24:467–500.

Lambert, D. M., and H. G. Spencer, eds. 1995. *Speciation and the Recognition Concept: Theory and Application*. Baltimore: Johns Hopkins University Press.

Tudge, C. 2000. *The Variety of Life: A Survey and Celebration of All the Creatures that Have Ever Lived*. Oxford: Oxford University Press.

Turner, A. 1995. Plio-Pleistocene correlations between climatic change and evolution in terrestrial mammals: The 2.5 Ma event in Africa and Europe. *Acta Zoologica Cracoviensia* 38:45–58.

Turner, A. 1999. Evolution in the African Plio-Pleistocene mammalian fauna: Correlation and causation. In T. G. Bromage and F. Schrenk, eds., *African Biogeography, Climate Change, and Human Evolution*, pp. 76–87. New York: Oxford University Press.

Turner, A., and H. E. H Paterson. 1991. Species and speciation: Evolutionary tempo and mode in the fossil record reconsidered. *Geobios* 24:761–769.

Turner, A., and B. A. Wood. 1993. Comparative palaeontological context for the evolution of the early hominid masticatory system. *Journal of Human Evolution* 24:301–318.

Vrba, E. S. 1985. Environment and evolution: Alternative causes of temporal distribution of evolutionary events. *South African Journal of Science* 81:229–236.

Vrba, E. S. 1993. Turnover-pulses, the Red Queen, and related topics. *American Journal of Science* 293-A:418–452.

Vrba, E. S. 1995. On the connections between paleoclimate and evolution. In E. S. Vrba, G. H. Denton, T. C. Partridge, and L. H. Burckle, eds., *Paleoclimate and Evolution, with Emphasis on Hominid Origins*, pp. 24–45. New Haven, Conn.: Yale University Press.

Vrba, E. S. 1999. Habitat theory in relation to the evolution in African Neogene biota and hominids. In T. G. Bromage and F. Schrenk, eds., *African Biogeography, Climate Change, and Human Evolution*, pp. 19–34. New York: Oxford University Press.

Vrba, E. S., G. H. Denton, and M. L. Prentice. 1989. Climatic influences on early hominid behaviour. *Ossa* 14:127–156.

Wesselman, H. B. 1984. *The Omo Micromammals: Systematics and Paleoecology of Early Man Sites from Ethiopia*. Contributions to Vertebrate Evolution, vol. 7. Basel: Karger.

DATING

Feibel, C. S., F. H. Brown, and I. McDougall. 1989. Stratigraphic context of fossil hominids from the Omo Group deposits: Northern Turkana Basin, Kenya and Ethiopia. *American Journal of Physical Anthropology* 78:595–622.

Smart, P. L., and P. D. Francis, eds. 1991. *Quaternary Dating Methods—A User's Guide*. Cambridge: Quaternary Research Association.

PAST AND PRESENT PHYSICAL GEOGRAPHY OF AFRICA

Adams, W. M., A. S. Goudie, and A. R. Orme, eds. 1996. *The Physical Geography of Africa*. Oxford: Oxford University Press.

Brown, F. H., and C. S. Feibel. 1988. "Robust" hominids and Plio-Pleistocene paleogeography of the Turkana Basin, Kenya. In F. E. Grine, ed., *The Evolutionary History of the "Robust" Australopithecines*, pp. 325–341. New York: Aldine de Gruyter.

Brown, F. H., and C. S. Feibel. 1991. Stratigraphy, depositional environments and palaeogeography of the Koobi Fora Formation. In J. M. Harris, ed., *Koobi Fora Research Project*. Vol. 3, *The Fossil Ungulates: Geology, Fossil Artiodactyls, and Palaeoenvironments*, pp. 1–30. Oxford: Clarendon Press.

Cooke, H. B. S. 1978. Africa: The physical setting. In V. J. Maglio and H. B. S. Cooke, eds., *The Evolution of African Mammals*, pp. 17–45. Cambridge, Mass.: Harvard University Press.

Feibel, C. S., J. M. Harris, and F. H. Brown. 1991. Palaeoenvironmental context for the late Neogene of the Turkana Basin. In J. M. Harris, ed., *Koobi Fora Research Project*. Vol. 3, *The Fossil Ungulates: Geology, Fossil Artiodactyls, and Palaeoenvironments*, pp. 321–370. Oxford: Clarendon Press.

Hofman, C., V. Courtillot, G. Féraud, P. Rochette, G. Yirgus, E. Ketefo, and R. Pik. 1997. Timing of the Ethiopian flood basalt event and implications for plume birth and global change. *Nature* 389:838–841.

Kingdon, J. 1989. *Island Africa: The Evolution of Africa's Rare Animals and Plants*. Princeton, N.J.: Princeton University Press.

Kirjksman, W., F. J. Hilgen, I. Raffi, F. J. Sierros, and D. S. Wilson. 1999. Chronology, causes and progression of the Messinian salinity crisis. *Nature* 400:652–655.

Pritchard, J. M. 1979. *Landform and Landscape in Africa*. London: Arnold.

Retallack, G. J. 1991. *Miocene Paleosols and Ape Habitats of Pakistan and Kenya*. Oxford: Oxford University Press.

Rögl, F. 1999. Mediterranean and Paratethys palaeogeography during the Oligocene and Miocene. In J. Agustí, L Rook, and P. Andrews, eds., *The Evolution of Terrestrial Ecosystems in Europe*, pp. 8–22. Cambridge: Cambridge University Press.

PAST AND PRESENT BIOGEOGRAPHY OF AFRICA

Behrensmeyer, A. K., N. E. Todd, R. Potts, and G. E. McBrinn. 1997. Late Pliocene faunal turnover in the Turkana Basin, Kenya and Ethiopia. *Science* 278:1589–1594.

Bromage, T. G., and F. Schrenk. 1995. Biogeographic and climatic basis for a narrative of early hominid evolution. *Journal of Human Evolution* 28:109–114.

Cox, C. B., and P. D. Moore. 2000. *Biogeography*. Oxford: Blackwell Scientific.

Grubb, P., O. Sandrock, O. Kullmar, T. M. Kaiser, and F. Schrenk. 1999. Relationships between eastern and southern African mammal faunas. In T. G. Bromage and F. Schrenk, eds., *African Biogeography, Climate Change, and Human Evolution*, pp. 253–267. New York: Oxford University Press.

Turner, A., and B. A. Wood. 1993. Taxonomic and geographic diversity in robust australopithecines and other African Plio-Pleistocene mammals. *Journal of Human Evolution* 24:147–168.

Turpie, J. K., and T. M. Crowe. 1994. Patterns of distribution, diversity and endemism of larger African mammals. *South African Journal of Zoology* 29:19–32.

Werger, M. J. A., ed. 1978. *Biogeography and Ecology of Southern Africa*. The Hague: Junk.

FOSSIL MAMMALS
General Treatments

Benton, M. J. 2000. *Vertebrate Palaeontology*. London: Chapman & Hall.

Carroll, R. L. 1988. *Vertebrate Paleontology and Evolution*. New York: Freeman.

Pough, F. H., C. M. Janis, and J. B. Heiser. 2001. *Vertebrate Life*. 6th ed. Upper Saddle River, N.J.: Prentice Hall.

Predators

Cooke, H. B. S. 1991. *Dinofelis barlowi* (Mammalia, Carnivora, Felidae) cranial material from Bolt's Farm, collected by the University of California African expeditions. *Palaeontologia Africana* 28:9–21.

Geraads, D. 1997. Carnivores du Pliocène terminal de Ahl al Oughlam (Casablanca, Maroc). *Geobios* 30:127–164.

Hendey, Q. B. 1974. The late Cenozoic Carnivora of the south-western Cape Province. *Annals of the South African Museum* 63:1–369.

Macdonald, D. 1992. *The Velvet Claw: A Natural History of the Carnivores*. London: BBC Books.

Morales, J., M. Pickford, D. Soria, and S. Fraile. 1998. New carnivores from the basal Middle Miocene of Arrisdrift, Namibia. *Ecologae Geologicae Helvetiae* 91:27–40.

Morales, J., M. J. Salesa, M. Pickford, and D. Soria. 2001. A new tribe, new genus and two new species of Barbourofelinae (Felidae, Carnivora, Mammalia) from the Early Miocene of East Africa and Spain. *Transactions of the Royal Society of Edinburgh: Earth Sciences* 92:97–102.

Savage, R. J. G. 1965. Fossil mammals of Africa: 19. The Miocene Carnivora of East Africa. *Bulletin of the British Museum (Natural History) Geology* 10:241–316.

Savage, R. G. J. 1973. *Megistotherium*, gigantic hyaenodont from Miocene of Gebel Zelten, Libya. *Bulletin of the British Museum (Natural History) Geology* 22:485–511.

Turner, A. 1990. The evolution of the guild of larger terrestrial carnivores during the Plio-Pleistocene in Africa. *Geobios* 23:349–368.

Turner, A. 1997. Further remains of Carnivora (Mammalia) from the Sterkfontein hominid site. *Palaeontologia Africana* 34:115–126.

Turner, A., and M. Antón. 1997. *Big Cats and Their Fossil Relatives: An Illustrated Guide to Their Evolution and Natural History*. New York: Columbia University Press.

Turner, A., and M. Antón. 1999. Climate and evolution: Implications of some extinction patterns in African and European machairodontine cats of the Plio-Pleistocene. *Estudios Geológicos* 54:209–230.

Werdelin, L. 1996. Carnivoran ecomorphology: A phylogenetic perspective. In J. L. Gittleman, ed., *Carnivore Behavior, Ecology, and Evolution*, pp. 582–624. Ithaca, N.Y.: Cornell University Press.

Werdelin, L., and M. E. Lewis. 2001. A revision of the genus *Dinofelis* (Mammalia, Felidae). *Zoological Journal of the Linnean Society* 132:147–258.

Werdelin, L., and N. Solounias. 1991. *The Hyaenidae: Taxonomy, Systematics and Evolution*. Fossils and Strata, no. 30. Oslo: Universitetsforlaget.

Werdelin, L., and A. Turner. 1996. The fossil and living Hyaenidae of Africa: Present status. In K. Stewart and K. Seymour, eds., *The Palaeoecology and Palaeoenvironments of Late Cenozoic Mammals: Tributes to the Career of C. S. Churcher*, pp. 637–659. Toronto: University of Toronto Press.

Werdelin, L., A. Turner, and N. Solounias. 1994. Studies of fossil hyaenids: The genera *Hyaenictis* Gaudry and *Chasmaporthetes* Hay, with a reconsideration of the Hyaenidae of Langebaanweg, South Africa. *Zoological Journal of the Linnean Society* 111:197–217.

Herbivores

Beden, M. 1983. Family Elephantidae. In J. M. Harris, ed., *Koobi Fora Research Project*. Vol. 2, *The Fossil Ungulates: Proboscidea, Perissodactyla, and Suidae*, pp. 40–129. Oxford: Clarendon Press

Beden, M. 1985. Les proboscidiens des grands gisements à hominidés Plio-Pléistocènes d'Afrique orientale. In Y. Coppens, ed., *L'Environnement des Hominidés au Plio-Pléistocène*, pp. 21–44. Paris: Masson.

Bernor, R. L., and M. Armour-Chelu. 1999. Towards an evolutionary history of African hipparionine horses. In T. G. Bromage and F. Schrenk, eds., *African Biogeography, Climate Change, and Human Evolution*, pp. 189–215. New York: Oxford University Press.

Bishop, L. C. 1999. Suid paleoecology and habitat preferences at African Pliocene and Pleistocene hominid localities. In T. G. Bromage and F. Schrenk, eds., *African Biogeography, Climate Change, and Human Evolution*, pp. 216–225. New York: Oxford University Press.

Eisenmann, V. 1983. Family Equidae. In J. M. Harris, ed., *Koobi Fora Research Project*. Vol. 2, *The Fossil Ungulates: Proboscidea, Perissodactyla, and Suidae*, pp. 156–214. Oxford: Clarendon Press.

Gentry, A. W. 1985. The Bovidae of the Omo Group deposits. In Y. Coppens and F. C. Howell, eds., *Les Faunes Plio-Pléistocènes de la Basse Vallée de L'Omo (Éthiopie)*. Vol. 1, *Perissodactyles, Artiodactyles (Bovidea)*, pp. 119–191. Paris: Centre National de la Recherche Scientifique.

Gentry, A. W. 1985. Pliocene and Pleistocene Bovidae in Africa. In Y. Coppens, ed., *L'Environnement des Hominidés au Plio-Pléistocène*, pp. 119–132. Paris: Masson.

Gèze, R. 1985. Répartition paléoécologique et relations phylogénéetiques des Hippopotamidae (Mammalia, Artiodactyla) du Néogène d'Afrique orientale. In Y. Coppens, ed., *L'Environnement des Hominidés au Plio-Pléistocène*, pp. 81–100. Paris: Masson.

Harris, J. M. 1983. Family Rhinocerotidae. In J. M. Harris, ed., *Koobi Fora Research Project*. Vol. 2, *The Fossil Ungulates: Proboscidea, Perissodactyla, and Suidae*, pp. 130–155. Oxford: Clarendon Press.

Harris, J. M. 1983. Family Suidae. In J. M. Harris, ed., *Koobi Fora Research Project*. Vol. 2, *The Fossil Ungulates: Proboscidea, Perissodactyla, and Suidae*, pp. 215–302. Oxford: Clarendon Press.

Harris, J. M. 1991. Family Bovidae. In J. M. Harris, ed., *Koobi Fora Research Project*. Vol. 3, *The Fossil Ungulates: Geology, Fossil Artiodactyls, and Palaeoenvironments*, pp. 139–320. Oxford: Clarendon Press.

Harris, J. M. 1991. Family Hippopotamidae. In J. M. Harris, ed., *Koobi Fora Research Project*. Vol. 3, *The Fossil Ungulates: Geology, Fossil Artiodactyls, and Palaeoenvironments*, pp. 31–85. Oxford: Clarendon Press.

Kalb, J. E., and A. Mebrate. 1993. Fossil elephantoids from the hominid-bearing Awash Group, Middle Awash Valley, Afar Depression, Ethiopia. *Transactions of the American Philosophical Society* 83:1–114.

Lindsay, E. H., N. D. Opdyke, and N. M. Johnson. 1980. Pliocene dispersal of the horse, *Equus*, and late Cenozoic mammalian dispersal events. *Nature* 287:135–138.

Vrba, E. S. 1985. African Bovidae: Evolutionary events since the Miocene. *South African Journal of Science* 81:263–266.

Vrba, E. S. 1995. The fossil record of African antelopes (Mammalia, Bovidae) in relation to human evolution and paleoclimate. In E. S. Vrba, G. H. Denton, T. C. Partridge, and L. H. Burckle, eds., *Paleoclimate and Evolution, with Emphasis on Hominid Origins*, pp. 385–424. New Haven, Conn.: Yale University Press.

Vrba, E. S. 1997. New fossils of Alcelaphini and Caprinae (Bovidae: Mammalia) from Awash, Ethiopia, and phylogenetic analysis of Alcelaphini. *Palaeontologia Africana* 34:127–198.

Primates

Andrews, P. 1992. Evolution and environment in the Hominoidea. *Nature* 360:641–646.

Asfaw, B., T. White, O. Lovejoy, B. Latimer, S. Simpson, and G. Suwa. 1999. *Australopithecus garhi*: A new species of early hominid from Ethiopia. *Science* 284:629–635.

Benefit, B. R. 1999. Biogeography, dietary specialization, and the diversification of African Plio-Pleistocene monkeys. In T. G. Bromage and F. Schrenk, eds., *African Biogeography, Climate Change, and Human Evolution*, pp. 172–188. New York: Oxford University Press.

Bromage, T. G., and F. Schrenk. 1995. Biogeographic and climatic basis for a narrative of early hominid evolution. *Journal of Human Evolution* 28:109–114.

Brunet, M. 2002. *Sahelanthropus* or "*Sahelpithecus*"?—A reply. *Nature* 419:582.

Brunet, M., A. Beauvilain, Y. Coppens, E. Heintz, A. H. E. Moutaye, and D. Pilbeam. 1996. *Australopithecus bahrelghazali*, a new species of early hominid from Koro Toro region, Chad. *Comptes Rendus de l'Académie des Sciences*, 2nd ser., 322:907–913.

Brunet, M., F. Guy, D. Pilbeam, H. T. Mackaye, A. Likius, D. Ahounta, A. Beauvillain, C. Blondel, H. Bocherens, J. R. Boisserie, L. de Bonis, Y. Coppens, J. Dejax, C. Denys, P. Duringer, V. Eisenmann, G. Fanone, P. Fronty, D. Geraads, T. Lehmann, F. Lihoreau, A. Louchart, A. Mahamat, G. Merceron, G. Mouchelin, O. Otero, P. P. Campomanes, M. Ponce de Leon, J. C. Rage, M. Sapanet, M. Schuster, J. Sudre, P. Tassy, X. Valentin, P. Vignaud, L. Viriot, A. Zazzo, and C. Zollikofer. 2002. A new hominid from the Upper Miocene of Chad, Central Africa. *Nature* 418:145–151.

Clark, J. D., Y. Beyene, G. WoldeGabriel, W. H. Hart, P. R. Renne, H. Gilbert, A. Defleur, G. Suwa, S. Katoh, K. R. Ludwig, J.-R. Boisserie, B. Asfaw, and T. D. White. 2003. Stratigraphic, chronological and behavioural contacts of Pleistocene *Homo sapiens* from Middle Awash, Ethiopia. *Nature* 423:747–752.

Halle-Selassie, Y. 2001. Late Miocene hominids from the Middle Awash, Ethiopia. *Nature* 412:178–181.

Jablonski, N. G., ed. 1993. *Theropithecus: The Rise and Fall of a Primate Genus*. Cambridge: Cambridge University Press.

Keyser, A. W. 2000. The Drimolen skull: The most complete australopithecine cranium to date. *South African Journal of Science* 96:189–193.

Leakey, M. G., C. S. Feibel, I. McDougall, and A. Walker. 1995. New four-million-year-old hominid species from Kanapoi and Allia Bay, Kenya. *Nature* 376:565–571.

Leakey, M. G., F. Spoor, F. H. Brown, P. N. Gathogo, C. Kiarie, L. N. Leakey, and I. McDougall. 2001. New hominin genus from eastern Africa shows diverse middle Pliocene lineages. *Nature* 410:433–440.

Senut, B., M. Pickford, D. Gommery, P. Mein, K. Cheboi, and Y. Coppens. 2001. First hominid from the Miocene (Lukeino Formation, Kenya). *Comptes Rendus de l'Académie des Sciences, Sciences de la Terre et des Planètes* 332:137–144.

Turner, A. 1999. Assessing earliest human settlement of Eurasia: Late Pliocene dispersions from Africa. *Antiquity* 73:563–570.

Vekua, A., D. Lordkipanidze, G. P. Rightmire, J. Agustí, R. Ferring, G. Maisuradze, A. Mouskhelishivili, M. Nioradze, M. Ponce de Leon, M. Tappen, M. Tvalchrelidze, and C. Zollikofer. 2002. A new skull of early *Homo* from Dmanisi, Georgia. *Science* 297:85–89.

Walker, A., and R. Leakey, eds. 1993. *The Nariokotome* Homo erectus *skeleton*. Berlin: Springer.

Ward, C. V., M. G. Leakey, and A. Walker. 2001. Morphology of *Australopithecus anamensis* from Kanapoi and Allia Bay, Kenya. *Journal of Human Evolution* 41:255–368.

Ward, S., B. Brown, A. Hill, J. Kelly, and W. Downs. 1999. *Equatorius*: A new hominoid genus from the Middle Miocene of Kenya. *Science* 285:1382–1386.

White, T. D., B. Asfaw, D. DeGusta, H. Tilbert, G. D. Richards, G. Suwa, and F. C. Howell. 2003. Pleistocene *Homo sapiens* from Middle Awash, Ethiopia. *Nature* 423:742–747.

White, T. D., G. Suwa, and B. Asfaw. 1994. *Australopithecus ramidus*, a new species of early hominid from Aramis, Ethiopia. *Nature* 371:306–312.

White, T. D., G. Suwa, and B. Asfaw. 1995. *Australopithecus ramidus*, a new species of early hominid from Aramis, Ethiopia: Corrigendum. *Nature* 375:88.

Wolpoff, M., B. Senut, M. Pickford, and J. Hawks. 2002. *Sahelanthropus* or "*Sahelpithecus*"? *Nature* 419:581–582.

Wood, B. A. 1991. *Koobi Fora Research Project*. Vol. 4, *Hominid Cranial Remains*. Oxford: Clarendon Press.

Wood, B. A. 1992. Origin and evolution of the genus *Homo*. *Nature* 355:783–790.

Wood, B. A., and M. Collard. 1999. The human genus. *Science* 284:65–71.

Fossil Faunas and Overviews

Agustí, J., and M. Antón. 2002. *Mammoths, Sabertooths, and Hominids: 65 Million Years of Mammalian Evolution in Europe*. New York: Columbia University Press.

Agustí, J., L. Rook, and P. Andrews, eds. 1999. *Hominoid Evolution and Climatic Change in Europe.* Vol. 1, *The Evolution of Neogene Terrestrial Ecosystems in Europe*. Cambridge: Cambridge University Press.

Bernor, L., V. Fahlbusch, and H.-W. Mittman, eds. 1996. *The Evolution of Western Eurasian Neogene Mammal Faunas.* New York: Columbia University Press.

Bromage, T. G., F. Schrenk, and Y. M. Juwayeyi. 1995. Paleobiogeography of the Malawi Rift: Age and vertebrate paleontology of the Chiwondo Beds, northern Malawi. *Journal of Human Evolution* 28:37–57.

Harris, J. M., and M. G. Leakey, eds. 2003. *Geology and Vertebrate Paleontology of the Early Pliocene Site of Kanapoi, Northern Kenya*. Contributions in Science, vol. 498. Los Angeles: Natural History Museum of Los Angeles County.

Maglio, V. J. 1975. Pleistocene faunal evolution in Africa and Eurasia. In K. W. Butzer and G. L. Isaac, eds., *After the Australopithecines*, pp. 419–476. The Hague: Mouton.

Maglio, V. J., and H. B. S. Cooke, eds. 1978. *The Evolution of African Mammals*. Cambridge, Mass.: Harvard University Press.

Owen-Smith, N. 1987. Pleistocene extinctions: The pivotal role of megaherbivores. *Paleobiology* 13:351–362.

Turner, A., L. Bishop, C. Denys, and J. McKee. 1999. A locality-based listing of African Plio-Pleistocene mammals. In T. G. Bromage and F. Schrenk, eds., *African Biogeography, Climate Change, and Human Evolution*, pp. 369–399. New York: Oxford University Press.

Whybrow, P. J., and A. Hill, eds. 1999. *Fossil Vertebrates of Arabia*. New Haven, Conn.: Yale University Press.

Reconstructions

Antón, M. 1997. Reconstrución de carnívoros fósiles. In R. García-Perea, R. A. Baquero, R. Fernández-Salvador, and J. Gisbert, eds. *Carnívoros: Evolución Ecología y Conservación*, pp. 137–152. Madrid: Consejo Superior de Investigaciones Científicas.

Antón, M., R. Garcia-Perea, and A. Turner. 1998. Reconstructed facial appearance of the sabre-toothed felid *Smilodon*. *Zoological Journal of the Linnean Society* 124:369–386.

Living Mammals

Bailey, T. N. 1993. *The African Leopard: Ecology and Behavior of a Solitary Felid*. New York: Columbia University Press.

Caro, T. M. 1994. *Cheetahs of the Serengeti Plains: Group Living in an Asocial Species*. Chicago: University of Chicago Press.

Estes, R. D. 1992. *The Behavior Guide to African Mammals*. Berkeley: University of California Press.

Greenacre, M. J., and E. S. Vrba. 1984. Graphical display and interpretation of antelope census data in African wildlife areas using correspondence analysis. *Ecology* 65:984–997.

Haltenorth, T., and H. Diller. 1980. *A Field Guide to the Mammals of Africa*. London: Collins.

Kingdon, J. 1997. *The Kingdon Field Guide to African Mammals*. San Diego, Calif.: Academic Press.

Kowalski, K., and B. Rzebik-Kowalska. 1991. *Mammals of Algeria*. Warsaw: Polish Academy of Sciences.

Kruuk, H. 1972. *The Spotted Hyena: A Study of Predation and Social Behavior*. Chicago: University of Chicago Press.

Meester, J. A. J., I. L. Rautenbach, N. J. Dippenaar, and C. M. Baker. 1986. *Classification of Southern African Mammals*. Monograph, no. 5. Pretoria: Transvaal Museum.

Mills, M. G. L. 1990. *Kalahari Hyaenas: The Comparative Ecology of Two Species*. London: Unwin Hyman.

Mills, M. G. L., and H. C. Biggs. 1993. Prey apportionment and related ecological relationships between large carnivores in Kruger National Park. *Symposium of the Zoological Society of London* 65:253–268.

Nowak, R. M. 1999. *Walker's Mammals of the World*. 6th ed. Baltimore: Johns Hopkins University Press.

Owen-Smith, N. 1985. Niche separation among African ungulates. In E. S. Vrba, ed., *Species and Speciation*, pp. 167–171. Monograph, no. 4. Pretoria: Transvaal Museum.

Piennar, U. deV. 1969. Predator–prey relationships amongst the larger mammals of the Kruger National Park. *Koedoe* 12:108–176.

Rautenbach, I. L. 1982. *Mammals of the Transvaal*. Monograph, no. 1. Pretoria: Ecoplan.

Schaller, G. B. 1972. *The Serengeti Lion: A Study of Predator–Prey Relations*. Chicago: University of Chicago Press.

Simmons, R. E., and L. Scheepers. 1996. Winning by a neck: Sexual selection in the evolution of the giraffe. *American Naturalist* 148:771–786.

Skinner, J. D., and R. H. N. Smithers. 1990. *The Mammals of the Southern African Subregion*. 2nd ed. Pretoria: University of Pretoria.

Wilson, D. E., and D. M. Reeder, eds. 1993. *Mammal Species of the World: A Taxonomic and Geographic Reference*. 2nd ed. Washington, D.C.: Smithsonian Institution Press.

Sites

Andrews, P. J. 1989. Paleoecology of Laetoli. *Journal of Human Evolution* 18:173–181.

Asfaw, B., Y. Beyene, G. Suwa, R. C. Walter, T. D. White, G. WoldeGabriel, and T. Yemane. 1992. The earliest Acheulean from Konso-Gardula. *Nature* 360:732–735.

Berger, L. R., and R. J. Clarke. 1995. Eagle involvement in accumulation of the Taung child fauna. *Journal of Human Evolution* 29:275–299.

Boaz, N. T., A. El-Arnauti, A. W. Gaziry, J. de Heinzelin, and D. D. Boaz, eds. 1987. *Neogene Paleontology and Geology of Sahabi*. New York: Liss.

Bonnefille, R., A. Vincens, and G. Buchet. 1987. Palynology, stratigraphy and palaeoenvironment of a Pliocene hominid site (2.9–3.3 M.Y.) at Hadar, Ethiopia. *Palaeogeography, Palaeoclimatology, Palaeoecology* 60:249–281.

Brain, C. K. 1981. *The Hunters or the Hunted?* Chicago: University of Chicago Press.

Brain, C. K., ed. 1993. *Swartkrans: A Cave's Chronicle of Early Man*. Monograph, no. 8. Pretoria: Transvaal Museum Monograph.

Bromage, T. G., and F. Schrenk, eds. 1995. *Evolutionary History of the Malawi Rift. Journal of Human Evolution* 28 [special issue]:37–57.

Brunet, M., A. Beauvilain, Y. Coppens, E. Heintz, A. H. E. Moutaye, and D. Pilbeam. 1995. The first australopithecine 2,500 kilometres west of the Rift Valley (Chad). *Nature* 378:273–275.

Bye, B. A., F. H. Brown, T. E. Cerling, and I. McDougall. 1987. Increased age estimate for the Lower Palaeolithic hominid site at Olorgesailie, Kenya. *Nature* 329:237–239.

Cadman, A., and R. J. Rayner. 1989. Climatic change and the appearance of *Australopithecus africanus* in the Makapansgat sediments. *Journal of Human Evolution* 18:107–113.

de Heinzelin, J., J. D. Clark, T. White, W. Hart, P. Renne, G. WoldeGabriel, Y. Beyene, and E. Vrba. 1999. Environment and behaviour of 2.5-million-year-old Bouri hominids. *Science* 284:625–629.

Gabunia, L., A. Vekua, D. Lordkipanidze, C. C. Swisher, R. Ferring, A. Justus, M. Nioradze, M. Tvalchrelidze, S. C. Anton, G. Bosinski, O. Jöris, M. A. de Lumley, G. Majsuradze, and A. Mouskhelishvili. 2000. Earliest Pleistocene hominid cranial remains from Dmanisi, Republic of Georgia: Taxonomy, geological setting, and age. *Science* 288:1019–1025.

Geraads, D., J.-J. Hublin, J.-J. Jaeger, H. Tong, S. Sen, and P. Toubeau. 1986. The Pleistocene hominid site of Ternifine, Algeria: New results on the environment, age and human industries. *Quaternary Research* 25:380–386.

Harris, J. M., F. H. Brown, and M. G. Leakey. 1988. *Stratigraphy and Paleontology of Pliocene and Pleistocene Localities West of Lake Turkana, Kenya.* Contributions in Science, vol. 399. Los Angeles: Natural History Museum of Los Angeles County.

Kappleman, J. 1991. The paleoenvironment of *Kenyapithecus* at Fort Ternan. *Journal of Human Evolution* 20:92–129.

Keyser, A. W., C. G. Menter, J. Moggi-Cecchi, T. R. Pickering, and L. R. Berger. 2000. Drimolen: A new hominid-bearing site in Gauteng, South Africa. *South African Journal of Science* 96:193–197.

Leakey, M. G., C. S. Feibel, R. L. Bernor, J. M. Harris, T. E. Cerling, K. M. Stewart, G. W. Storrs, A. Walker, L. Werdelin, and A. J. Winkler. 1996. Lothagam: A record of faunal change in the late Miocene of East Africa. *Journal of Vertebrate Paleontology* 16:556–570.

Leakey, M. G., and J. M. Harris, eds. 1987. *Laetoli: A Pliocene Site in Northern Tanzania.* Oxford: Clarendon Press.

Leakey, M. G., and J. M. Harris, eds. 2003. *Lothagam: The Dawn of Humanity in Eastern Africa.* New York: Columbia University Press.

Menter, C. G., K. L. Kuykendall, A. W. Keyser, and G. C. Conroy. 1999. First record of hominid teeth from the Plio-Pleistocene site of Gondolin, South Africa. *Journal of Human Evolution* 37:299–307.

Pickford, M., and B. Senut. 2001. The geological and faunal context of Late Miocene hominid remains from Lukeino, Kenya. *Comptes Rendus de l'Académie des Sciences, Sciences de la Terre et des Planètes* 332:145–152.

Savage, R. J. G., and W. R. Hamilton. 1973. Introduction to the Miocene mammal faunas of Gebel Zelten, Libya. *Bulletin of the British Museum (Natural History)* 22:516–527.

Turner, A. 1989. Sample selection, schelpp effects and scavenging: The implications of partial recovery for interpretations of the terrestrial mammal assemblage from Klasies River Mouth. *Journal of Archaeological Science* 16:1–11.

Vignaud, P., P. Duringer, H. T. Mackaye, A. Likius, C. Blondel, J. R. Boisserie, L. de Bonis, V. Eisenmann, M. E. Etienne, D. Geraads, F. Guy, T. Lehmann, F. Lihoreau, N. Lopez-Martinez, C. Mourer-Chauvire, O. Otero, J. C. Rage, M. Schuster, L. Viriot, A. Zazzo, and M. Brunet. 2002. Geology and palaeontology of the Upper Miocene Toros-Menalla hominid locality, Chad. *Nature* 418:152–155.

Watson, V. 1993. Glimpses from Gondolin: A faunal analysis of a fossil site near Broederstroom, Transvaal, South Africa. *Palaeontologia Africana* 30:35–42.

WoldeGabriel, G., Y. Haile-Selassie, P. R. Tenne, W. K. Hart, S. H. Ambrose, B. Asfaw, G. Heiken, and T. White. 2001. Geology and palaeontology of the Late Miocene Middle Awash Valley, Afar Rift, Ethiopia. *Nature* 412:175–178.

WoldGabriel, G., T. D. White, G. Suwa, P. Renne, J. de Heinzelin, and W. K. Heiken. 1994. Ecological and temporal placement of early Pliocene hominids at Aramis, Ethiopia. *Nature* 371:330–333.

Taxonomy

Madsen, O., M. Scally, C. J. Douady, D. J. Kao, R. W. DeBry, R. Adkins, H. M. Amrine, M. J. Stan-
hope, W. W. de Jong, and M. S. Springer. 2001. Parallel adaptive radiations in two major clades
of placental mammals. *Nature* 409:610–614.
Murphy, W. J., E. Eizirick, W. E. Johnson, Y. P. Zhang, O. A. Ryder, and S. J. O'Brien. 2001. Molecular phylogenetics and the origins of placental mammals. *Nature* 409:614–618.

Vegetation and Habitat, Past and Present

Andrews, P. 1996. Palaeoecology and hominoid palaeoenvironments. *Biological Reviews*
71:257–300.
Behrensmeyer, A. K., J. D. Damuth, W. A. DiMichele, R. Potts, H. D. Sues, and S. L. Wing, eds.
1992. *Terrestrial Ecosystems Through Time.* Chicago: University of Chicago Press.
Bonnefille, R. 1985. Evolution of the continental vegetation: The palaeobotanical record from East
Africa. *South African Journal of Science* 81:267–270.
Bonnefille, R. 1995. A reassessment of the Plio-Pleistocene pollen record of East Africa. In
E. S. Vrba, G. H. Denton, T. C. Partridge, and L. H. Burckle, eds., *Paleoclimate and Evolution,
with Emphasis on Hominid Origins*, pp. 299–310. New Haven, Conn.: Yale University Press.
Cerling, T. E. 1992. Development of grasslands and savannas in East Africa during the Neogene.
Palaeogeography, Palaeoclimatology, Palaeoecology 97:241–247.
Cerling, T. E., J. M. Harris, B. J. MacFadden, M. G. Leakey, J. Quade, V. Eisenmann, and J. R. Ehle-
ringer. 1997. Global vegetation change through the Miocene/Pliocene boundary. *Nature*
389:153–158.
Hamilton, A. C. 1982. *Environmental History of East Africa.* London: Academic Press.
Martin, C. 1991. *The Rain Forests of West Africa: Ecology-Threats-Conservation.* Basel: Bikhäuser.
McClanahan, T. R., and T. P. Young, eds. 1966. *East African Ecosystems and Their Conservation.*
New York: Oxford University Press.
O'Brien, E. M., and C. R. Peters. 1999. Landforms, climate, ecogeographic mosaics, and the poten-
tial for hominid diversity in Pliocene Africa. In T. G. Bromage and F. Schrenk, eds., *African
Biogeography, Climate Change, and Human Evolution*, pp. 115–137. New York: Oxford University
Press.
Owen-Smith, N. 1999. Ecological links between African savanna environments, climate change,
and early hominid evolution. In T. G. Bromage and F. Schrenk, eds., *African Biogeography,
Climate Change, and Human Evolution*, pp. 138–149. New York: Oxford University Press.
Sikes, N. E. 1994. Early hominid habitat preferences in East Africa: Paleosol carbon isotopic evi-
dence. *Journal of Human Evolution* 27:25–45.
Sikes, N. E. 1999. Plio-Pleistocene floral context and habitat preferences of sympatric hominid
species in East Africa. In T. G. Bromage and F. Schrenk, eds., *African Biogeography, Climate
Change, and Human Evolution*, pp. 301–315. New York: Oxford University Press.

INDEX

Numbers *in italics refer to pages on which figures appear.*